AF361829

MEMS Inertial Device

MEMS Inertial Device

Editor

Huiliang Cao

Basel • Beijing • Wuhan • Barcelona • Belgrade • Novi Sad • Cluj • Manchester

Editor
Huiliang Cao
School of Instrument and
Electronics
North University of China
Taiyuan
China

Editorial Office
MDPI
St. Alban-Anlage 66
4052 Basel, Switzerland

This is a reprint of articles from the Special Issue published online in the open access journal *Micromachines* (ISSN 2072-666X) (available at: www.mdpi.com/journal/micromachines/special_issues/0WG5V77K21).

For citation purposes, cite each article independently as indicated on the article page online and as indicated below:

Lastname, A.A.; Lastname, B.B. Article Title. *Journal Name* **Year**, *Volume Number*, Page Range.

ISBN 978-3-7258-0018-6 (Hbk)
ISBN 978-3-7258-0017-9 (PDF)
doi.org/10.3390/books978-3-7258-0017-9

Contents

micromachines

MDPI

Editorial

Editorial for the Special Issue on Micro-Electromechanical System Inertial Devices

Huiliang Cao

State Key Laboratory of Dynamic Measurement Technology, North University of China, Taiyuan 030051, China; caohuiliang@nuc.edu.cn

Citation: Cao, H. Editorial for the Special Issue on Micro-Electromechanical System Inertial Devices. *Micromachines* **2023**, *14*, 2134. https://doi.org/10.3390/mi14122134

Received: 13 October 2023
Accepted: 15 November 2023
Published: 21 November 2023

Micro-electromechanical systems (MEMS) are miniature systems comprising micro-mechanical sensors, actuators, and microelectronic circuits [1]. With the explosive growth of core technologies such as MEMS miniaturization, microelectronic integration, and high-precision mass production [2], various types of MEMS sensors have found wide-ranging applications in fields such as aerospace [3], petrochemicals [4], the marine and automotive industries [5], household appliances [6], and healthcare [7]. In this context, MEMS inertial devices represent a core category of MEMS sensors. They are designed to capture physical motion, such as linear displacement or angular rotation, and convert these responses into electrical signals, which are subsequently amplified and processed through electronic circuits [8]. MEMS inertial devices primarily include MEMS gyroscopes [9], MEMS accelerometers [10], MEMS magnetometers [11], and MEMS-IMUs [12]. Accelerometers and gyroscopes are the most common MEMS inertial sensors. The former are sensitive to axial acceleration and convert it into usable output signals [13], while gyroscopes detect the angular velocity of a moving body relative to inertial space [14]. Furthermore, the combination of three MEMS accelerometers and three MEMS gyroscopes forms a micro-inertial measurement unit (MIMU) capable of sensing linear acceleration in three directions and angular acceleration in three directions [15]. In summary, MEMS inertial microsystems employ three-dimensional heterogeneous integration technology, integrating MEMS accelerometers, gyroscopes, pressure sensors, magnetic sensors, signal processing circuits, and embedded algorithms into silicon chips to achieve chip-level guidance, navigation, positioning, and other functionalities [16].

In terms of applications, MEMS inertial devices are mainly used to provide accurate position and motion measurement solutions for aerospace, underwater exploration, robot control and many other core fields. Take NASA's Mars rover as an example, where MEMS accelerometers and gyroscopes enable the rover to land and navigate precisely on the Martian surface. These sensors are, therefore, essential for space exploration and can ensure accurate data collection and telemetry [17]. In the field of underwater exploration, MEMS inertial sensors play an important role in navigation for autonomous underwater vehicles in complex underwater terrain. The REMUS AUV, developed by Woods Hole Oceanographic Institution, uses MEMS inertial sensors to map the seabed and conduct scientific research [18]. The integration of MEMS inertial sensors into robots has enhanced the capabilities of robots in industrial automation. Boston Dynamics' four-legged robot Spot employs MEMS sensors for balance and stability, allowing it to traverse uneven terrain with amazing agility and precision [19]. In the field of automatic driving, MEMS inertial sensors are an indispensable component for safe and efficient navigation. Tesla and other companies use MEMS-based inertial measurement units (IMUs) to provide accurate real-time data for their automatic driving system, ensuring accurate control of vehicle movement and improving the safety and reliability of automated driving technology [20]. The emergence of wearable devices has brought MEMS inertial sensors closer to our daily lives. Fitness trackers such as Fitbit utilize MEMS accelerometers to monitor physical activity and track steps taken by the user, providing valuable health insights [21].

On the other hand, MEMS inertial devices also face a number of challenges, such as low measurement accuracy due to temperature, noise, and their own limitations. They are deployed in extremely variable working environments, facing high temperatures, high pressures, high inertia levels, and high impacts, resulting in demanding requirements for the stability and adaptability of MEMS inertial devices [22]. In order to fully develop and utilize the potential of these devices for high-end manufacturing and cutting-edge applications, researchers and engineers must study their structural optimization, measurement and control systems, manufacturing technologies, and integration applications. This Special Issue presents the latest advances in MEMS inertial devices with the purpose of encouraging readers to explore each article.

This Special Issue contains 11 papers covering different inertial devices such as MEMS accelerometers (Contributions 1–5), MEMS gyroscopes (Contributions 6–10), and inertial navigation systems (Contribution 11). Half of these papers discuss the design and manufacture of inertial devices, such as anchor stress relief (Contribution 1), modal simulation, and structural optimization (Contributions 3 and 4), closed-loop drive circuit design (Contribution 7), and the development of pattern-matching closed-loop systems (Contribution 10). The other half of the papers focus on the optimization and maintenance of MEMS inertial devices, including noise removal (Contribution 2), hardware compensation (Contribution 5), algorithm compensation (Contribution 6), and fault diagnosis (Contribution 8). Additionally, the effects of the resonator's coupling efficiency on the scale factor of the fiber resonator gyroscope is discussed in (Contribution 9), providing a theoretical reference and experimental basis for various applications on land, sea, and air. The accuracy of inertial units in navigation and positioning systems is discussed in (Contribution 11).

In particular, Liu et al. (Contribution 1) improved the performance of an all-silicon accelerometer by adjusting the ratio of the Si-SiO_2 bonding area and the Au-Si bonding area in the anchor zone to eliminate the stress. Their results show that the zero-bias full-temperature stability and scale-factor full-temperature stability can be improved significantly. Cui et al. (Contribution 3) manufactured a three-pole-plate dual-capacitance acceleration sensor and used COMSOL, ANSYS and other software to simulate and optimize its structure and shock characteristics. They obtained a sensor with self-powered output, a high output voltage amplitude, and low spurious interference, which can reliably receive vibration signals. Shi et al. (Contribution 4) proposed a piezoelectric MEMS accelerometer (PMA) with four cantilever beams integrated with inertial mass elements to meet the requirement of high-sensitivity acceleration for vector hydrophones. They established a theoretical model for energy harvesting for a piezoelectric cantilever beam, and the geometric size and structure of their micro-device are optimized to meet the vibration absorption conditions. Han et al. (Contribution 7) analyzed the problem of automatic gain control driving a low-Q micromachined gyroscope at room temperature and atmospheric pressure. They proposed a drive circuit based on frequency modulation, which uses the second harmonic demodulation circuit to eliminate the same frequency coupling between the drive signal and the displacement signal. Wu et al. (Contribution 10) proposed a VCF-based mode-matching micromachine-optimized tuning fork gyroscope, which can maximize the scale factor and avoid the use of additional orthogonal rings. On this basis they established a mode-matching, closed-loop system without quadrature-nulling loop, and the corresponding convergence and matching error were quantitatively analyzed.

In the optimization and maintenance of inertial devices, Wang et al. (Contribution 2) proposed a hybrid algorithm based on empirical mode decomposition (EMD) and time–frequency peak filtering (TFPF) to deal with the noise during accelerometer calibration. Faced with the problem of temperature drift, Liu et al. (Contribution 5) designed a combined compensation method using reference voltage source compensation and accelerometer terminal temperature compensation, based on the idea of hardware compensation. They comprehensively improved the performance of the accelerometer within a wide temperature range. Li et al. (Contribution 6) adopted the idea of algorithmic compensation and combined empirical mode decomposition (EMD), a radial basis function neural network (RBF NN), a genetic algorithm (GA) and the Kalman filter (KF) to propose a new fusion algorithm to remove the influence of environment and accurately compensate the temperature drift of a MEMS gyroscope. In order to achieve reliable maintenance, Cui et al. (Contribution 8) designed a dual-mass MEMS gyroscope fault diagnosis platform, which integrates the Simulink structure model of the gyroscope and the measurement and control system, and reserves a variety of algorithm interfaces for users to independently program. This can effectively identify and classify the seven signals of the gyroscope: normal, bias, blocking, drift, multiplicity, cycle, and internal fault. On the applications side, Li et al. (Contribution 9) evaluated the effect of the coupling efficiency of fiber ring resonators on scale factors. This provides a theoretical reference and experimental basis for various applications at sea, on land, and in space. In order to improve the positioning accuracy of shearers, Lu et al. (Contribution 11) established an experimental ground shearer installation based on the positioning model of a tri-INS and discussed the influence of inertial navigation system installation parameters on positioning accuracy.

I would like to take this opportunity to thank all the authors for submitting their papers for this special issue. I would also like to thank all the reviewers for dedicating their time and helping to improve the quality of the submitted papers.

Conflicts of Interest: The authors declare no conflict of interest.

List of Contributions

1. Liu, G.; Liu, Y.; Ma, X.; Wang, X.; Zheng, X.; Jin, Z. Research on a Method to Improve the Temperature Performance of an All-Silicon Accelerometer. *Micromachines* **2023**, *14*, 869. https://doi.org/10.3390/mi14040869.
2. Wang, C.; Cui, Y.; Liu, Y.; Li, K.; Shen, C. High-G MEMS Accelerometer Calibration Denoising Method Based on EMD and Time-Frequency Peak Filtering. *Micromachines* **2023**, *14*, 970. https://doi.org/10.3390/mi14050970.
3. Cui, M.; Chuai, S.; Huang, Y.; Liu, Y.; Li, J. Structural Design of MEMS Acceleration Sensor Based on PZT Plate Capacitance Detection. *Micromachines* **2023**, *14*, 1565. https://doi.org/10.3390/mi14081565.
4. Shi, S.; Ma, L.; Kang, K.; Zhu, J.; Hu, J.; Ma, H.; Pang, Y.; Wang, Z. High-Sensitivity Piezoelectric MEMS Accelerometer for Vector Hydrophones. *Micromachines* **2023**, *14*, 1598. https://doi.org/10.3390/mi14081598.
5. Liu, G.; Liu, Y.; Li, Z.; Ma, Z.; Ma, X.; Wang, X.; Zheng, X.; Jin, Z. Combined Temperature Compensation Method for Closed-Loop Microelectromechanical System Capacitive Accelerometer. *Micromachines* **2023**, *14*, 1623. https://doi.org/10.3390/mi14081623.
6. Li, Z.; Cui, Y.; Gu, Y.; Wang, G.; Yang, J.; Chen, K.; Cao, H. Temperature Drift Compensation for Four-Mass Vibration MEMS Gyroscope Based on EMD and Hybrid Filtering Fusion Method. *Micromachines* **2023**, *14*, 971. https://doi.org/10.3390/mi14050971.
7. Han, T.; Wang, G.; Dong, C.; Jiang, X.; Ren, M.; Zhang, Z. A Self-Oscillating Driving Circuit for Low-Q MEMS Vibratory Gyroscopes. *Micromachines* **2023**, *14*, 1057. https://doi.org/10.3390/mi14051057.
8. Cui, R.; Ma, T.; Zhang, W.; Zhang, M.; Chang, L.; Wang, Z.; Xu, J.; Wei, W.; Cao, H. A New Dual-Mass MEMS Gyroscope Fault Diagnosis Platform. *Micromachines* **2023**, *14*, 1177. https://doi.org/10.3390/mi14061177.
9. Li, S.; Tian, X.; Tian, S. Research on Optical Fiber Ring Resonator Q Value and Coupling Efficiency Optimization. *Micromachines* **2023**, *14*, 1680. https://doi.org/10.3390/mi14091680.
10. Wu, Y.; Yuan, W.; Xue, Y.; Chang, H.; Shen, Q. Virtual Coriolis-Force-Based Mode-Matching Micromachine-Optimized Tuning Fork Gyroscope without a Quadrature-Nulling Loop. *Micromachines* **2023**, *14*, 1704. https://doi.org/10.3390/mi14091704.
11. Lu, C.; Wang, S.; Shin, K.; Dong, W.; Li, W. Experimental Research of Triple Inertial Navigation System Shearer Positioning. *Micromachines* **2023**, *14*, 1474. https://doi.org/10.3390/mi14071474.

References

1. de Groot, W.A.; Webster, J.R.; Felnhofer, D.; Gusev, E.P. Review of Device and Reliability Physics of Dielectrics in Electrostatically Driven MEMS Devices. *IEEE Trans. Device Mater. Reliab.* **2009**, *9*, 190–202. [CrossRef]
2. Zhu, J.; Liu, X.; Shi, Q.; He, T.; Sun, Z.; Guo, X.; Liu, W.; Sulaiman, O.B.; Dong, B.; Lee, C. Development Trends and Perspectives of Future Sensors and MEMS/NEMS. *Micromachines* **2020**, *11*, 7. [CrossRef]
3. Ghazali, M.H.M.; Rahiman, W. Fuzzy-Oriented Anomaly Inspection in Unmanned Aerial Vehicle (UAV) Based on MEMS Accelerometers in Multimode Environment. *IEEE Trans. Instrum. Meas.* **2023**, *72*, 3530710. [CrossRef]

4. Bhat, K.P.; Oh, K.W.; Hopkins, D.C. Feasibility of a MEMS Sensor for Gas Detection in HV Oil-Insulated Transformer. *IEEE Trans. Ind. Appl.* **2013**, *49*, 316–321. [CrossRef]
5. Zhang, W.; Hao, C.; Zhang, Z.; Yang, S.; Peng, J.; Wu, B.; Xue, X.; Zang, J.; Chen, X.; Yang, H.; et al. Vector High-Resolution Marine Turbulence Sensor Based on a MEMS Bionic Cilium-Shaped Structure. *IEEE Sens. J.* **2021**, *21*, 8741–8750. [CrossRef]
6. KMori; Misawa, K.; Ihida, S.; Takahashi, T.; Fujita, H.; Toshiyoshi, H. A MEMS Electrostatic Roll-Up Window Shade Array for House Energy Management System. *IEEE Photonics Technol. Lett.* **2016**, *28*, 593–596. [CrossRef]
7. JBaik; Seo, S.; Lee, S.; Yang, S.; Park, S.-M. Circular Radio-Frequency Electrode with MEMS Temperature Sensors for Laparoscopic Renal Sympathetic Denervation. *IEEE Trans. Biomed. Eng.* **2022**, *69*, 256–264. [CrossRef]
8. Sabato, A.; Niezrecki, C.; Fortino, G. Wireless MEMS-Based Accelerometer Sensor Boards for Structural Vibration Monitoring: A Review. *IEEE Sens. J.* **2017**, *17*, 226–235. [CrossRef]
9. Marx, M.; Cuignet, X.; Nessler, S.; De Dorigo, D.; Manoli, Y. An Automatic MEMS Gyroscope Mode Matching Circuit Based on Noise Observation. *IEEE Trans. Circuits Syst. II Express Briefs* **2019**, *66*, 743–747. [CrossRef]
10. Morichika, S.; Sekiya, H.; Zhu, Y.; Hirano, S.; Maruyama, O. Estimation of Displacement Response in Steel Plate Girder Bridge Using a Single MEMS Accelerometer. *IEEE Sens. J.* **2021**, *21*, 8204–8208. [CrossRef]
11. Perrier, T.; Levy, R.; Bourgeteau-Verlhac, B.; Kayser, P.; Moulin, J.; Paquay, S. Optimization of an MEMS Magnetic Thin Film Vibrating Magnetometer. *IEEE Trans. Magn.* **2017**, *53*, 4000705. [CrossRef]
12. Lu, J.; Ye, L.; Zhang, J.; Luo, W.; Liu, H. A New Calibration Method of MEMS IMU Plus FOG IMU. *IEEE Sens. J.* **2022**, *22*, 8728–8737. [CrossRef]
13. Zhang, H.; Sobreviela, G.; Pandit, M.; Chen, D.; Sun, J.; Parajuli, M.; Zhao, C.; Seshia, A.A. A Low-Noise High-Order Mode-Localized MEMS Accelerometer. *J. Microelectromechanical Syst.* **2021**, *30*, 178–180. [CrossRef]
14. Ren, J.; Zhou, T.; Zhou, Y.; Li, Y.; Su, Y. An In-Run Automatic Mode-Matching Method with Amplitude Correction and Phase Compensation for MEMS Disk Resonator Gyroscope. *IEEE Trans. Instrum. Meas.* **2023**, *72*, 7505911. [CrossRef]
15. Liu, S.Q.; Zhang, J.C.; Li, G.Z.; Zhu, R. A Wearable Flow-MIMU Device for Monitoring Human Dynamic Motion. *IEEE Trans. Neural Syst. Rehabil. Eng.* **2020**, *28*, 637–645. [CrossRef]
16. Chen, S.; Zhao, Q.; Cui, J. Effect of Proton Radiation on Mechanical Structure of Silicon MEMS Inertial Devices. *IEEE Trans. Electron Devices* **2022**, *69*, 5155–5161. [CrossRef]
17. NASA Jet Propulsion Laboratory. Mars Science Laboratory (Curiosity Rover). 2021. Available online: https://mars.jpl.nasa.gov/msl/ (accessed on 6 November 2023).
18. Woods Hole Oceanographic Institution. REMUS. 2021. Available online: https://www.whoi.edu/what-we-do/explore/vehicles/remus/ (accessed on 7 December 2021).
19. Boston Dynamics. Spot. 2021. Available online: https://www.bostondynamics.com/spot (accessed on 23 May 2021).
20. Tesla. Autopilot. 2021. Available online: https://www.tesla.com/autopilot (accessed on 28 June 2021).
21. Fitbit. Fitbit. 2021. Available online: https://www.fitbit.com/global/us/home (accessed on 12 August 2021).
22. Dean, R.N.; Castro, S.T.; Flowers, G.T.; Roth, G.; Ahmed, A.; Hodel, A.S.; Grantham, B.E.; Bittle, D.A.; Brunsch, J.P. A Characterization of the Performance of a MEMS Gyroscope in Acoustically Harsh Environments. *IEEE Trans. Ind. Electron.* **2011**, *58*, 2591–2596. [CrossRef]

micromachines

Article

A New Dual-Mass MEMS Gyroscope Fault Diagnosis Platform

Rang Cui [1,†], Tiancheng Ma [2,†], Wenjie Zhang [2,†], Min Zhang [3,*], Longkang Chang [4], Ziyuan Wang [5], Jingzehua Xu [2], Wei Wei [2] and Huiliang Cao [1,*]

[1] Key Laboratory of Instrumentation Science & Dynamic Measurement, Ministry of Education, North University of China, Taiyuan 030051, China; cuirang@outlook.com
[2] Tsinghua Shenzhen International Graduate School, Tsinghua University, Shenzhen 518055, China; mtc21@mails.tsinghua.edu.cn (T.M.); zhang-wj21@mails.tsinghua.edu.cn (W.Z.); 19955778426@163.com (J.X.); weiw20@mails.tsinghua.edu.cn (W.W.)
[3] School of Instrument Science and Opto-Electronics Engineering, Beijing Information Science & Technology University, Beijing 100192, China
[4] College of Intelligence Science and Technology, National University of Defense Technology, Changsha 410073, China; changlongkang1999@126.com
[5] Department of Electronic Engineering, Tsinghua University, Beijing 100084, China; wangziyu21@mails.tsinghua.edu.cn
[*] Correspondence: zhangmin_19990301@163.com (M.Z.); caohuiliang@nuc.edu.cn (H.C.)
[†] These authors contributed equally to this work.

Abstract: MEMS gyroscopes are one of the core components of inertial navigation systems. The maintenance of high reliability is critical for ensuring the stable operation of the gyroscope. Considering the production cost of gyroscopes and the inconvenience of obtaining a fault dataset, in this study, a self-feedback development framework is proposed, in which a dualmass MEMS gyroscope fault diagnosis platform is designed based on MATLAB/Simulink simulation, data feature extraction, and classification prediction algorithm and real data feedback verification. The platform integrates the dualmass MEMS gyroscope Simulink structure model and the measurement and control system, and reserves various algorithm interfaces for users to independently program, which can effectively identify and classify seven kinds of signals of the gyroscope: normal, bias, blocking, drift, multiplicity, cycle and internal fault. After feature extraction, six algorithms, ELM, SVM, KNN, NB, NN, and DTA, were respectively used for classification prediction. The ELM and SVM algorithms had the best effect, and the accuracy of the test set was up to 92.86%. Finally, the ELM algorithm is used to verify the actual drift fault dataset, and all of them are successfully identified.

Keywords: MEMS gyroscope; fault diagnosis platform; feature extraction

Citation: Cui, R.; Ma, T.; Zhang, W.; Zhang, M.; Chang, L.; Wang, Z.; Xu, J.; Wei, W.; Cao, H. A New Dual-Mass MEMS Gyroscope Fault Diagnosis Platform. *Micromachines* **2023**, *14*, 1177. https://doi.org/10.3390/mi14061177

Academic Editor: Nam-Trung Nguyen

Received: 6 May 2023
Revised: 28 May 2023
Accepted: 29 May 2023
Published: 31 May 2023

1. Introduction

As an important device in navigation systems, gyroscopes can provide accurate navigation positioning and attitude parameters for carriers, and their accuracy and reliability play a rather important role in navigation systems, and are used in various aspects of spaceflight, aviation, and navigation [1]. With the development of modern technology, measurement and control systems are becoming more and more complex, resulting in difficulties regarding troubleshooting and increased possibility of failure. To improve the accuracy and reliability of the gyroscope, promptly providing navigation and positioning parameters for the carrier, detecting, identifying, and predicting navigation system faults, and guaranteeing the positioning accuracy and reliability of the navigation system are important tasks [2].

A fault diagnosis system needs to be able to perform three tasks: fault detection, to determine whether a fault has occurred in the system; fault isolation, to locate the fault and determine which sensor or actuator in the system has failed; and fault identification, to estimate the size, type or characteristics of the fault [3].

There are presently three primary categories of fault diagnosis technique: mathematical model-based diagnosis, input/output signal analysis and processing-based diagnosis, and artificial intelligence-based diagnosis.

Mathematical model-based fault diagnosis was first proposed by Dr. Beard of MIT in his thesis in 1971, and requires the establishment of a more accurate mathematical model of the research object in order to study the dynamic characteristics of the object and reproduce the control process at the theoretical level through mathematical means, thus achieving a reproduction and diagnosis of faults [4]. However, sometimes the mathematical model of the system is difficult to establish precisely, leading to a certain instability of the final diagnosis. The latter two types of fault diagnosis method do not require an accurate mathematical model to be provided, and also overcome redundant dependencies, leading them to be a hot topic at present for research in the field of gyroscope fault detection and diagnosis.

Signal processing-based methods are used to obtain monitoring signals by monitoring the process of the system, and to find the intrinsic relationship between the signals by extracting certain eigenvalues in the monitoring signals for fault detection [5]. Common methods include Fourier transform, wavelet transforms [6], empirical modal decomposition [7], and other methods. In [8], wavelet packet transform was used to extract the features of the satellite control moment gyroscope. Pu et al. [9] studied the application of Local Mean Decomposition (LMD) in depth for the diagnosis of mechanical faults in high-voltage circuit breakers.

Artificial intelligence-based fault diagnosis is an emerging research direction at present. It mainly relies on research results from current hot topics in the field of artificial intelligence, including support vector machines (SVM), fault trees, fuzzy logic, rough sets, neural networks, etc. Zhang et al. [10] proposed a new fusion fault diagnosis method based on FFT-based SAE and WPD-based NN for the fault diagnosis of gyroscopes; Zhao et al. [11] put forward a new CNN network scheme based on attention-enhanced convolutional blocks (AECB) for a data-driven CMG fault diagnosis scheme.

At present, it is quite difficult to perform diagnosis of complex faults due to the cost of gyroscope fabrication and the inconvenience of fault dataset acquisition, so in this study, a self-feedback development framework is proposed to design a novel dualmass MEMS gyroscope fault diagnosis platform. The platform has the following advantages:

(1)   The platform is a complete self-feedback system, integrating a dualmass MEMS gyroscope Simulink structure model and measurement and control system, gyroscope fault signal simulation model, data feature extraction and identification classification algorithm, and real data feedback verification.

(2)   The platform can generate seven types of signal: normal, bias, blocking, drift, multiplicity, period, and internal fault, and those signals are identified and classified using six algorithms—ELM, SVM, KNN, NB, NN, and DTA—after feature extraction, all of which have good diagnosis and recognition rates.

(3)   Although the neural network model can generally achieve good diagnosis and recognition rate when applied to fault diagnosis, there are also some problems, so the platform also reserves various algorithm interfaces for users' independent programming, so that users can optimize the algorithm by themselves and connect the algorithm to this platform to verify the effectiveness of the algorithm.

2. Dualmass MEMS Gyroscope Fault Diagnosis Platform

Overall Approach

Figure 1 depicts a novel dualmass MEMS gyroscope fault diagnosis platform that is based on the self-feedback development framework proposed in this paper. As illustrated in the figure, the platform consists of three key components: the gyroscope model, the theoretical model, and the platform model. The first component involves the analysis and modeling of the dualmass MEMS gyroscope's structure, while the second component focuses on the analysis and modeling of the gyroscope's measurement and control system [12–14]. The third component, the platform model, is primarily responsible for generating, extracting features from, identifying, and classifying various types of gyroscope fault. Subsequently, feedback verification of the experimental data is conducted, and this section provides a detailed explanation of the theoretical foundations of each of these three components.

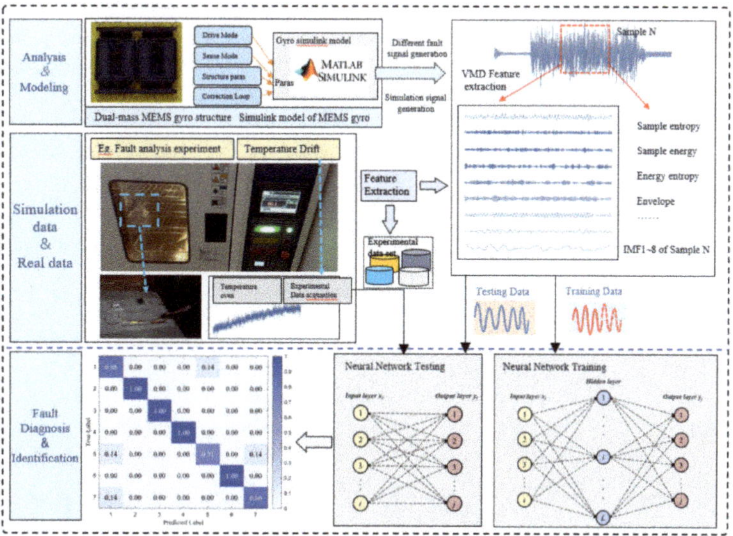

Figure 1. A new dualmass MEMS gyroscope fault diagnosis platform.

3. The Principle of Dualmass MEMS Gyroscopes

3.1. Structural Model of Dual Mass Silicon Micromechanical Gyroscope

Dualmass MEMS gyroscopes are mainly composed of mechanically sensitive structures of silicon materials and driving and detection circuits [15,16]. Of these, the silicon structure possess a driving mode and a detection mode, and can convert the angular rate input signal into the displacement signal of the detection mode through the Coriolis effect. The driving circuit provides the necessary vibration conditions for the driving mode, and the detection circuit extracts the Coriolis signal. The gyroscope structure used is fully decoupled, as shown in Figure 2. In the absence of external impact and vibration, the kinetic equation of the gyroscope is described as:

$$\begin{bmatrix} m_x & 0 \\ 0 & m_y \end{bmatrix} \begin{bmatrix} \ddot{x} \\ \ddot{y} \end{bmatrix} + \begin{bmatrix} c_{xx} & c_{xy} \\ c_{yx} & c_{yy} \end{bmatrix} \begin{bmatrix} \dot{x} \\ \dot{y} \end{bmatrix} + \begin{bmatrix} k_{xx} & k_{xy} \\ k_{yx} & k_{yy} \end{bmatrix} \begin{bmatrix} x \\ y \end{bmatrix} = \begin{bmatrix} F_{dx} + 2m_p\Omega_z\dot{y} \\ F_{dy} - 2m_p\Omega_z\dot{x} \end{bmatrix} \quad (1)$$

In the equation, m_x and m_y denote the equivalent quality of the drive mode and detection mode; c_{xx}, c_{yy}, k_{xx}, and k_{yy} are the effective damping and stiffness for the driving and detection modes, respectively; c_{xy} and k_{xy} are the damping and stiffness of the detection mode coupled to the driving mode; c_{yx} and k_{yx} are effective damping and stiffness for coupling the driving mode to the detection mode; F_{dx} is the driving force; F_{dy} is the feedback force of the detection mode; M_p denotes the Coriolis mass; Ω_z represents the angular rate around the z-axis; x and y represent the displacement in the driving and detection directions [17,18].

When the gyroscope structure is working, the driving frame will vibrate with constant frequency (resonant frequency of the driving mode) and amplitude along the x-direction. With an Ω_z input, the Coriolis mass will drive the detection frame to move along the y-axis under the Coriolis force (which is related to Ω_z). The mechanical thermal noise present in the gyroscope structure can be equivalent to a random, zero average Gaussian force, and its effect on the gyroscope can be equated to adding forces next to the dampers:

$$F_n = \sqrt{4K_B TCB} \tag{2}$$

where F_n is the equivalent interference force of mechanical thermal noise; $K_B = 1.38 \times 10^{-23}$ J/K is the Boltzmann constant; T is the absolute temperature; C and B is damping coefficient and working bandwidth of the gyroscope structure respectively. Due to the significant proportion of Coriolis mass in detecting equivalent mass, by substituting Equation (2) into Equation (1) with $m_y = m_p$, we can obtain:

$$\begin{bmatrix} m_x & 0 \\ 0 & m_y \end{bmatrix} \begin{bmatrix} \ddot{x} \\ \ddot{y} \end{bmatrix} + \begin{bmatrix} c_{xx} & c_{xy} \\ c_{yx} & c_{yy} \end{bmatrix} \begin{bmatrix} \dot{x} \\ \dot{y} \end{bmatrix} + \begin{bmatrix} k_{xx} & k_{xy} \\ k_{yx} & k_{yy} \end{bmatrix} \begin{bmatrix} x \\ y \end{bmatrix} = \begin{bmatrix} F_{dx} + F_{nx} + 2m_p\Omega_z\dot{y} \\ F_{dy} + F_{ny} - 2m_p\Omega_z\dot{x} \end{bmatrix} \tag{3}$$

From Equation (3), it can be analyzed that the driving mode is a second-order differential equation for forced vibration, and for the convenience of analysis, the driving displacement is set as:

$$x(t) = A_x \sin(\omega_d t) \tag{4}$$

where A_x is the vibration amplitude of the driving mode; ω_d is driving force frequency equal to ω_x.

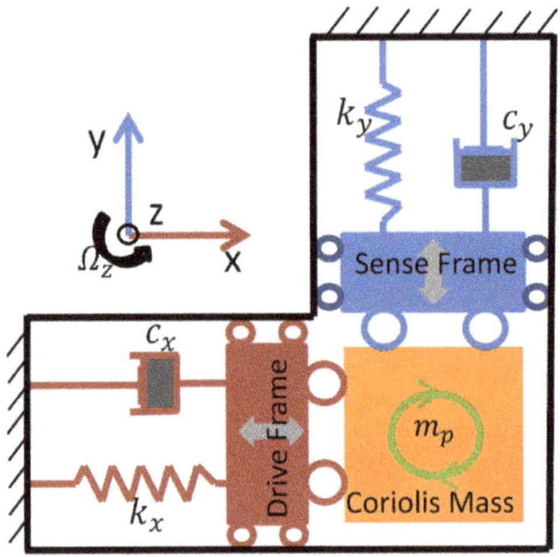

Figure 2. Schematic of fully decoupled structure.

3.2. Dualmass MEMS Gyroscope Monitoring System

To realize a series of measures to improve gyroscope performance, such as quadrature correction and the detection of a closed loop, in this paper, modules for achieving these functions are added to the gyroscope model, and the simulation model diagram is displayed in Figure 3. The model contains the drive mode, the detection mode, a mutual coupling module between the drive and detection modes, a mechanical thermal noise module (this paper adopts the value of room temperature, 20 °C), a quadrature correction module, and a Ghosh homogeneous correction force module.

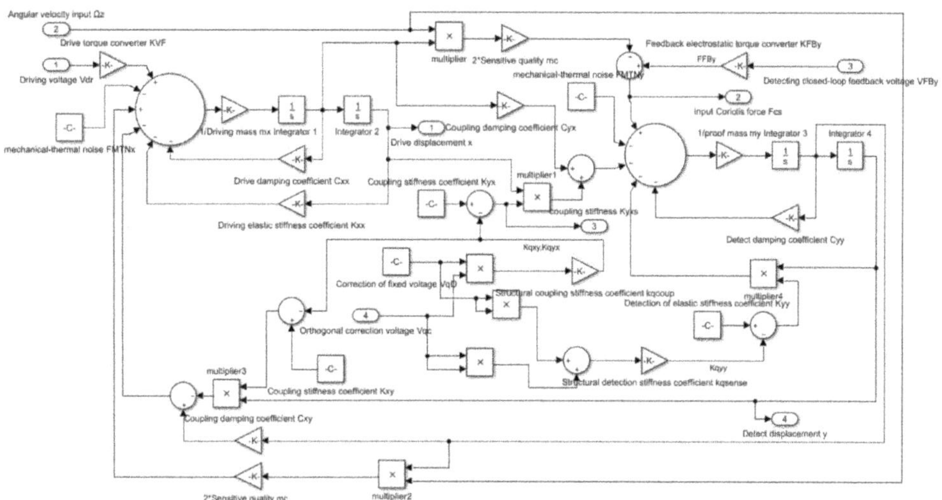

Figure 3. Equivalent model of gyroscope structure.

The model input signals are drive voltage V_d, angular velocity input Ω_z, detection feedback voltage V_{yb}, and quadrature correction voltage V_{fs}. The output signals include drive displacement signal x, detection displacement signal y, and detection modal force signal F_q.

If the resonant frequencies $f_x = \frac{1}{2\pi}\omega_x = \frac{1}{2\pi}\sqrt{\frac{k_{xx}}{m_x}}$ and $f_y = \frac{1}{2\pi}\omega_y = \frac{1}{2\pi}\sqrt{\frac{k_{yy}}{m_y}}$ of the driving mode and the detection mode are set at 4050 Hz and 4060 Hz, respectively, in the model, the operating bandwidth B is approximately 5.4 Hz. Simultaneously, the quality factors $Q_x = \frac{\sqrt{m_x k_{xx}}}{c_{xx}}$ and $Q_y = \frac{\sqrt{m_y k_{yy}}}{c_{yy}}$ of both modes are set to 2000, and it is assumed that the equivalent input angular velocities caused by coupling stiffness and damping are 200 (°)/s and 5 (°)/s, respectively. The calculation method for equivalent input angular velocity can be obtained from the following two equations:

$$\begin{cases} -2\Omega_k m_p \dot{x} = x k_{yx} \\ -2\Omega_c m_p \dot{x} = x c_{yx} \end{cases} \tag{5}$$

where Ω_k and Ω_C are the equivalent input angular rates caused by coupling stiffness and coupling damping, respectively. Since the gyroscope structure often adopts a symmetrical structure, this article assumes that $k_{yx} = k_{xy}$, $c_{yx} = c_{xy}$.

The conversion of voltage and force in the model is achieved through comb capacitors and satisfies the equation:

$$\begin{cases} F_{static} = \frac{1}{2}\nabla C(x,y,z)V_{static}^2 \\ C(x,y,z) = n\frac{\varepsilon h l}{d} \end{cases} \tag{6}$$

where F_{static} is the electrostatic force; $C(x,y,z)$ is the comb capacitance; V_{static} is the voltage generated by electrostatic force; n is the number of comb teeth; ε is the dielectric constant; h is the structural thickness; l is the superimposed length of the comb capacitor; d is the spacing between comb capacitors [19,20].

3.3. Platform Model

A Simulink simulation model platform based on the gyroscope model and theoretical model is built, as shown in Figure 4. In this platform, the gyroscope measurement and control system and the gyroscope fault simulation model are designed, and then feature extraction and processing algorithms are designed by MATLAB/Simulink. The established model for the entire system was used for fault signal recognition and classification. The gyroscope fault simulation model was used to generate bias, blocking, drift, multiplicity, period, and internal fault signals, combined with normal signals for feature extraction, and classification was performed using six algorithms: ELM, SVM, KNN, NB, NN, and DTA. At the same time, the platform also reserves algorithm recognition interfaces for users to independently program various algorithms into the platform to prove the effectiveness of algorithms.

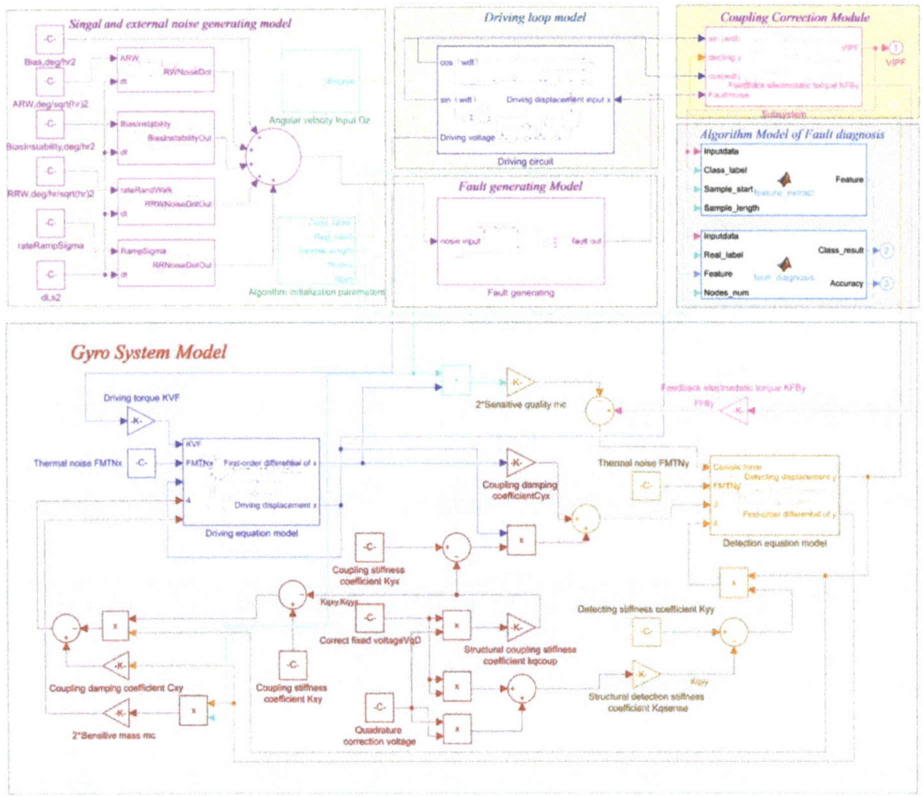

Figure 4. Simulink-based platform model.

4. Algorithms

4.1. Variational Mode Decomposition (VMD)

The original signal $f(t)$ was decomposed into K intrinsic mode components (IMF) via a constrained variational model, as shown in Equation (7):

$$\begin{cases} \min_{(u_k)(w_k)} \left\{ \sum_k \left\| \partial_t [\sigma(t) + \frac{i}{t} u_t(t)] e^{-jw_k t} \right\|_2^2 \right\} \\ s.t \sum_k u_k = f(t) \end{cases} \tag{7}$$

In this context, $\{u_k\} = \{u_1, u_2, \ldots, u_k\}$ are the k IMFs, $\{w_k\} = \{w_1, w_2, \ldots, w_k\}$ denotes a series of center frequencies that were utilized in the decomposition process. Additionally, σ_t represents the impulse function that was employed in the analysis.

The Lagrange operator $\lambda(t)$ and the quadratic penalty factor α are introduced to transform the inequality constraint into an equation constraint, and the corresponding extended Lagrange expressions can be written as:

$$L(\{u_k\}, \{w_k\}, \lambda) = \alpha \sum_{k=1}^{k} \left\| \partial_t \left[\left(\delta(t) + \frac{j}{\pi t} \right) \times u_k(t) \right] e^{iw_k t} \right\|_2^2 + \left\| f(t) - \sum_{k=1}^{k} u_k(t) \right\|_2^2 + \left\langle \lambda(t), f(t) - \sum_{k=1}^{k} u_k(t) \right\rangle \tag{8}$$

The optimal solution to Equation (7) is obtained through the use of an alternate multiplier method to locate the saddle point of Equation (8). The iterative updating process for u_k and ω_k is shown in Equations (9) and (10), respectively. The analysis involved refining the intrinsic mode components (IMFs) and center frequencies u_k and ω_k through an iterative process. By utilizing the alternate multiplier method, the optimal solution to Equation (7) was obtained.

$$\hat{u}_k^{n+1}(\omega) = \frac{\hat{f}(\omega) - \sum_{i<k} \hat{u}_i^{n+1}(\omega) - \sum_{i<k} \hat{u}_{i<k}^{n+1}(\omega) + \frac{\hat{\lambda}^n(\omega)}{2}}{1 + 2\alpha(\omega - \omega_k^n)^2} \tag{9}$$

$$\omega_k^{n+1} = \frac{\int_0^\infty \omega \left| \hat{u}_k^{n+1}(\omega) \right|^2 d\omega}{\int_0^\infty \left| \hat{u}_k^{n+1}(\omega) \right|^2 d\omega} \tag{10}$$

The value of $\lambda(\omega)$ was updated using the following equation:

$$\lambda^{n+1}(\omega) = \lambda^n(\omega) + \tau(f(\omega) - \sum_{k}^{n+1} u_k(w)) \tag{11}$$

The above steps were repeated until the iteration stop condition was reached [21]:

$$\sum_k \left\| u_k^{n+1} - u_k^n \right\|_2^2 / \left\| u_k^n \right\|_2^2 < \gamma \tag{12}$$

4.2. Sample Entropy (SE)

The size of sample entropy (SE) can be used to measure the self-similarity and complexity of the data sequence, and the calculation of SE does not depend on the data length of the signal, so it has better statistical stability and adaptability when quantifying the characteristics of the gyroscope fault signal [22]. The calculation steps of SE are as follows:

Step 1: Given the original time series $x = \{x_1, x_2, \ldots, x_N\}$, m is chosen as a suitable embedding dimension to construct a new state vector, which is $x = \{x_i, x_{i+1}, \ldots, x_{i+m-1}\}$, $i = 1, 2, \ldots, N - m$.

Step 2: The maximum difference distance between two state vectors is calculated as follows:

$$d(x_i, x_j) = \max_k \left(\left| x_{i+k} - x_{j+k} \right| \right), k = 0, 1, \ldots, m - 1 \tag{13}$$

Step 3: Given the similarity tolerance parameter r, B_i is calculated, whose distance between x_i and x_j is less than or equal to r, and defined as follows:

$$B_i^m(r) = \frac{B_i}{N - m - 1}, 1 \leq i \leq N - m \tag{14}$$

Step 4: Calculate the mean and define $B^m(r)$:

$$B^m(r) = \frac{1}{N - m + 1} \sum_{i=1}^{N-m+1} B_i^m(r) \tag{15}$$

Step 5: Increase the embedding dimension $m + 1$, and repeat the above four steps to calculate $B^{m+1}(r)$:

$$B^{m+1}(r) = \frac{1}{N-m} \sum_{i=1}^{N-m} B_i^{m+1}(r) \tag{16}$$

When the length of the time series is finite, SE can be calculated as follows:

$$SE(m,r,N) = -\ln\left(\frac{B^{m+1}(r)}{B^m(r)}\right) \tag{17}$$

4.3. Extreme Learning Machine (ELM)

The Extreme Learning Machine (ELM) represents a novel single-hidden-layer feedforward neural network architecture. One of its key features is the ability to set the parameters of hidden layer nodes either randomly or manually without the need for complex adjustments. Additionally, the learning process only involves a calculation of the output weight, significantly reducing computational complexity compared to other neural network models. By introducing a nonlinear activation function in the hidden layer, the ELM is capable of handling complex, nonlinear phenomena in an efficient and effective manner. Therefore, the convergence rate of the ELM algorithm is much faster than that of the traditional algorithm, because it does not require iteration. At the same time, random hidden nodes guarantee the global approximation ability [23]. Therefore, this paper proposes using ELM as a network model to predict the fault signal of the gyroscope. The network model of ELM is shown in Figure 5.

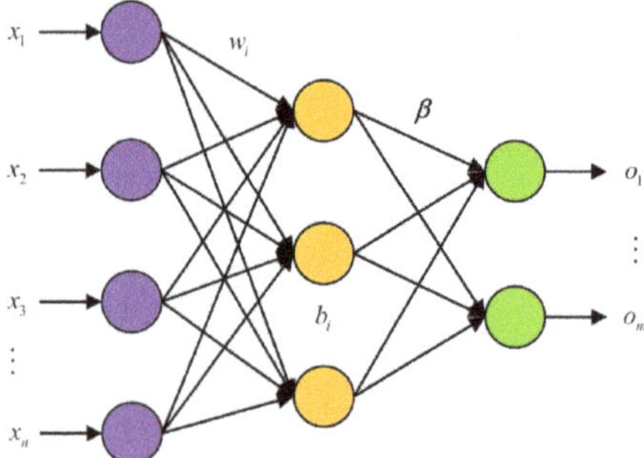

Figure 5. The network model of ELM.

For the ELM network model training phase, N different input/output (x_i, t_i) are needed, which the $x_i = [x_{i1}, x_{i2}, \ldots, x_{in}]^T \in R^n$, $t_i = [t_{i1}, t_{i2}, \ldots, t_{im}]^T \in R^m$. The ELM network model, which comprises L hidden layer nodes, can be represented as follows:

$$\sum_{i=1}^{L} \beta_i h(\omega_i \cdot x_j + b_i) = O_j, (j = 1 \cdots N) \tag{18}$$

where $h(x)$ is the hidden layer activation function; in ELM, the activation function provides nonlinear mapping for the system; $\omega_i = [\omega_{i1}, \omega_{i2}, \ldots, \omega_{in}]^T$ is the input weight matrix; β_i is the output weight matrix; b_i is the i-th hidden layer bias; $\omega_i \cdot x_j$ is the inner product of ω_i and x_j; O_j represents the output of the model.

The goal of ELM network model training is to reduce the discrepancy between the target and the ELM output. This can be expressed mathematically as:

$$\sum_{j=1}^{N} \| O_j - t_j \| = 0 \tag{19}$$

There are β_i, ω_i, and b_i, which make:

$$\sum_{i=1}^{L} \beta_i h(\omega_i \cdot x_j + b_i) = t_j, (j = 1 \cdots N) \tag{20}$$

The matrix can be expressed as:

$$H\beta = T \tag{21}$$

where the output of the hidden layer node is denoted by:

$$H = \begin{bmatrix} h(\omega_1 \cdot x_1 + b_1) & \cdots & h(\omega_L \cdot x_L + b_L) \\ \vdots & \cdots & \vdots \\ h(\omega_1 \cdot x_N + b_1) & \cdots & h(\omega_L \cdot x_N + b_L) \end{bmatrix}_{N \times L}$$

and the output weight matrix is represented by β and T is the target matrix.

Finally, the expression of its solution is as follows:

$$\hat{\beta} = \left(H^T H \right)^{-1} H^T T \tag{22}$$

4.4. SVM (Support Vector Machine)

Support vector machine (SVM) is a supervised machine learning algorithm based on the principle of minimum structural risk [24]. SVM can classify problems into linear and nonlinear types, to solve nonlinear problems, SVM introduces a kernel function $k(x,z)$, in which radial basis function can map original data into high-dimensional space, and the expression is:

$$k(x,z) = \exp\left(-\frac{||x-z||^2}{2\sigma^2}\right) \tag{23}$$

After introducing the kernel function, the SVM classification decision function is

$$g(x) = sign\left(\sum_{i=1}^{n} a_i y_i k(x_i, x) + b\right) \tag{24}$$

To solve the problem whereby some sample points are still inseparable after projection into high-dimensional space, the relaxation variable ξ_i is introduced into the classification model for optimization adjustment, and penalty factor C is introduced to the loss part. The new model is obtained by introducing the relaxation variable ξ_i and penalty factor C:

$$\begin{cases} \min_{w,b} \frac{1}{2} \| w \|^2 + C \sum_{i=1}^{n} \xi_i \\ y_i(w^T x_i + b) \geq 1 - \xi_i, \xi_i \geq 0, i = 1, 2, \cdots, n. \end{cases} \tag{25}$$

Convert it to a dual problem:

$$\begin{cases} w_{\max} = \sum_{i=1}^{n} a_i - \frac{1}{2} \sum_{i=1,j=1}^{n} a_i a_j y_i y_j k(x_i, x_j) \\ \text{s.t.} \sum_{i=1}^{n} a_i y_i = 0, 0 \leq a_i \leq C, i = 1, 2, \cdots, n \end{cases} \tag{26}$$

where a_i and a_j represent Lagrange multipliers; y_i and y_j represent the training sample category, where y_i and $y_j \in \{-1, 1\}$. Penalty factor C is used to measure the complexity of the learning machine: if the value of C is too large, overfitting is likely to occur, and the SVM model will tend to be complex, with longer running time and reduced operational efficiency; if the value of C is too small, the fitting degree of the data sample will be reduced, and the SVM model will be prone to underfitting [25].

5. Simulation and Verification of the Platform Model

Firstly, the signal of the gyroscope is obtained, and after obtaining the signal, the output is first decomposed using the VMD algorithm; after that, the signal is decomposed by sample entropy, sample energy, energy entropy, and envelope algorithm for different frequencies bands to obtain the signal features, which constitute the feature dataset; secondly, the training set and test set of feature samples are divided, and the training set is used for training, and the test evaluation is completed using the test set for training.

5.1. Simulation

Different types of faults may occur in the gyroscope during long-term operation due to different fault-inducing factors. According to different types, gyroscope faults can be divided into the following types:

Bias fault: this refers to the constant difference between the output signal and the actual value after a certain moment. Its mathematical description form is:

$$y_s = \begin{cases} y(t) & t < t_s \\ y(t) + d & t \geq t_s \end{cases} \tag{27}$$

Block fault: this refers to a moment when the output of the gyroscope is stuck at a constant bias, keeping its output value unchanged. Its mathematical model is:

$$y_s = \begin{cases} y(t) & t < t_s \\ y(t_s) & t \geq t_s \end{cases} \tag{28}$$

Drift fault: due to changes in the surrounding environment or internal parameters, such as temperature changes or calibration issues, the output of the gyroscope has an increased constant term. The output value often accumulates additional errors over time. Its mathematical expression is:

$$y_s = \begin{cases} y(t) & t < t_s \\ y(t) + kt & t \geq t_s \end{cases} \tag{29}$$

where k is the drift rate.

Multiplicative fault: this means that the output signal of the gyro is multiplied with a multiplication factor from a certain point in time, mainly due to scaling errors in the output. Its mathematical expression is:

$$y_s = \begin{cases} y(t) & t < t_s \\ ky(t) & t \geq t_s \end{cases} \tag{30}$$

Periodic fault: this means that the output signal of the gyroscope is attached to a signal of periodic change from a certain time. Its mathematical expression is:

$$y_s = \begin{cases} y(t) & t < t_s \\ y(t) + square(t) & t \geq t_s \end{cases} \tag{31}$$

where $square(t)$ is the square wave signal, which is expressed as:

$$square(t) = square(t + T_0) \tag{32}$$

$$square(t) = \begin{cases} d & 0 \leq t < \frac{T_0}{2} \\ -d & \frac{T_0}{2} < t < T_0 \end{cases} \tag{33}$$

Internal structure fault: the equivalent mass, stiffness, and damping of the gyroscope have changed.

To verify the effectiveness of the proposed dualmass MEMS gyroscope fault diagnosis platform, we conducted simulations to verify it. The Simulink simulation platform was used to generate seven types of signal: normal signals and six fault types of signals, as shown in Figures 6 and 7, where Figure 6 indicates the output form of normal signals and Figure 7 indicates the output form of fault signals. Sixty-seven sets of signals were generated, of which forty sets were used to train the model and twenty-seven sets were used to test the accuracy of the model.

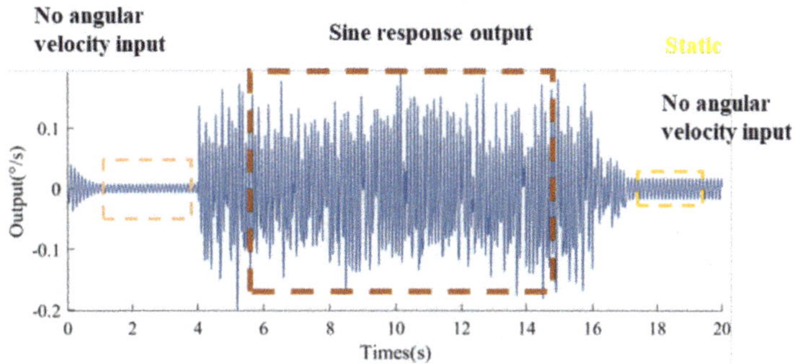

Figure 6. Normal signal of MEMS gyroscope.

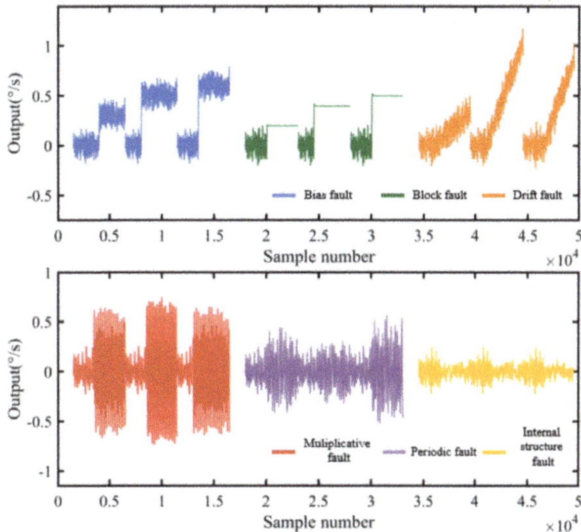

Figure 7. Fault class of the dualmass MEMS gyroscope.

Feature extraction was performed for seven different signals, and classification was performed using six algorithms—ELM [26], SVM [27], KNN [28], NB [29], NN [30], and DTA [31]—and the recognition rate of the classification results is shown in Figure 8, from which we can see that the accuracy of classification recognition using the ELM algorithm was 92.86%, which was able to identify 100% of bias, blocking, drift, period, and internal fault signals, respectively. The accuracy of classification recognition using the SVM algorithm was the same as that of ELM algorithm; the accuracy of classification recognition using the KNN algorithm was 89.29%, whereby it was able to identify 100% of bias, blocking, drift, period and internal fault signals, 50% of normal signals and 75% of multiplicative signals. The accuracy of classification recognition using the NB algorithm was 89.29%, whereby bias, blocking, drift and period signals were identified with 100% probability, and normal, multiplicative and internal fault signals were identified with 75% probability. The accuracy of classification recognition using the NN algorithm was 85.71%, in which bias, blocking, drift and period signals were identified with 100% probability, while 50% of the normal signals were recognized, 75% of the multiplicative signals and 75% of the internal fault signals. Classification recognition using the DTA algorithm was 82.14%, whereby 100% of the bias, blocking, drift and periodic signals, 25% of the normal signals, 75% of the multiplicative signals and 75% of the internal fault signals were recognized. To show the accuracy of identification more visually, histograms were obtained, as shown in Figure 9. From Figure 9, it can be seen that both ELM and SVM have high classification accuracy compared to other fault diagnosis algorithms. The combined analysis shows that the best results were obtained using ELM with the SVM algorithm, and the diagnosis results show the effectiveness and accuracy of the method.

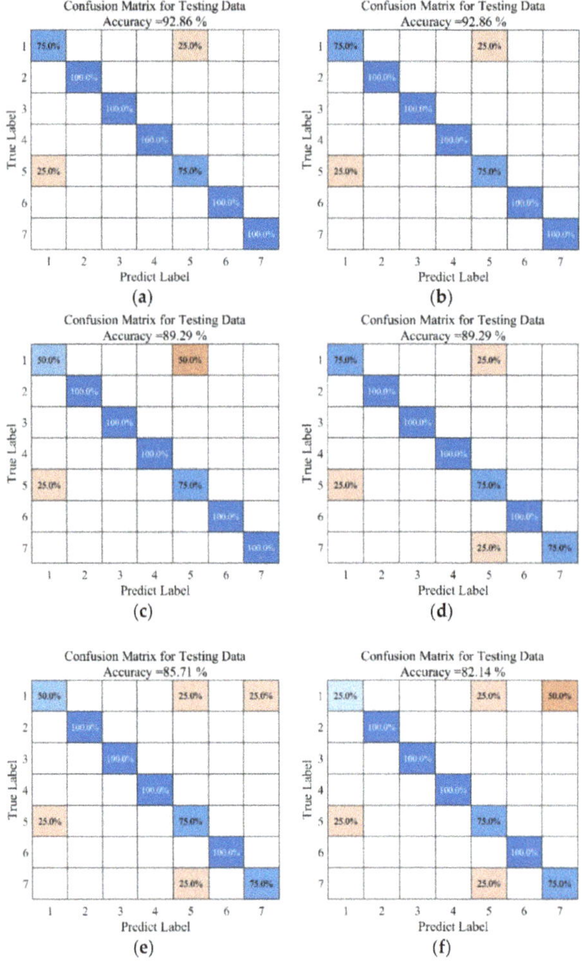

Figure 8. The recognition results of different algorithms: (**a**) ELM; (**b**) SVM; (**c**) KNN; (**d**) NB; (**e**) NN; (**f**) DTA.

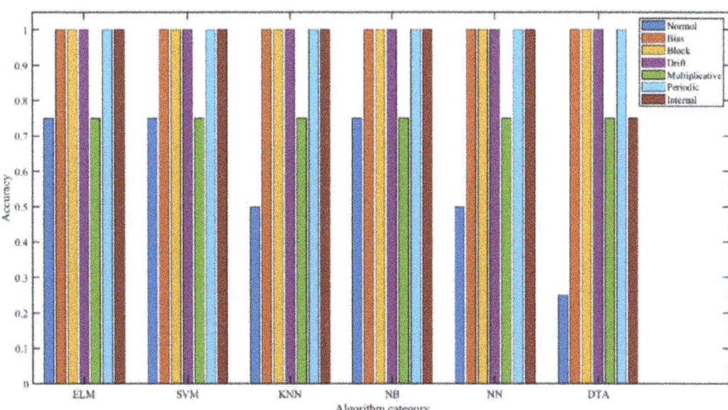

Figure 9. Bar chart of accuracy with seven groups for cross-validation.

For the trained model, to verify the effectiveness of the training, twenty-eight groups of signals were used for verification in this paper, and the seven types of signals, namely, normal, bias, blocking, drift, multiplicative, periodic, and internal fault, are labeled as the corresponding numbers [1–7]. Finally, the prediction diagram of the fault diagnosis platform obtained is shown in Figure 10, where the blue dots indicate the actual signal type, and the black lines indicate the recognized signal type. From the figure, it can be seen that only the 16th and 23rd sample point fault types were not accurately identified, while the rest were accurately identified, and the recognition rate of the fault diagnosis platform proposed in this paper is as high as 92.9%, which can effectively and accurately identify the type of signal. Therefore, it can be said that both the training classification accuracy and the test classification accuracy of the fault diagnosis model can reach an equal or higher classification level.

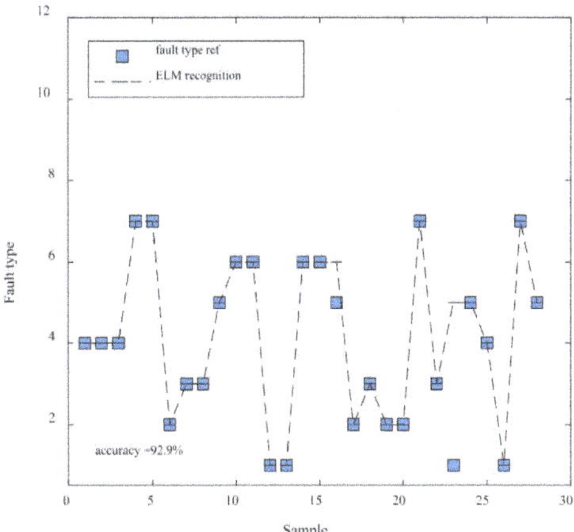

Figure 10. Test results.

5.2. Experiment Verification

To assess the validity and efficacy of the MEMS gyroscope fault diagnosis platform, we conducted a temperature experiment to obtain drift signal data from an actual gyroscope. Figure 11 displays both the gyroscope and the experimental equipment employed in this study. Specifically, we utilized the dual mass-07# gyroscope for our experiment, with detection circuits arranged across three separate PCB boards and connected through the use of metal pins to transmit electronic signals and maintain structural integrity. To minimize electromagnetic interference, the metal case was grounded to the "GND" signal. The first of the three PCB boards served as an interface for processing weak signals, with the structural chip connected to it. The remaining two PCB boards were utilized for the sensing loop and the drive closed loop, respectively [32–34].

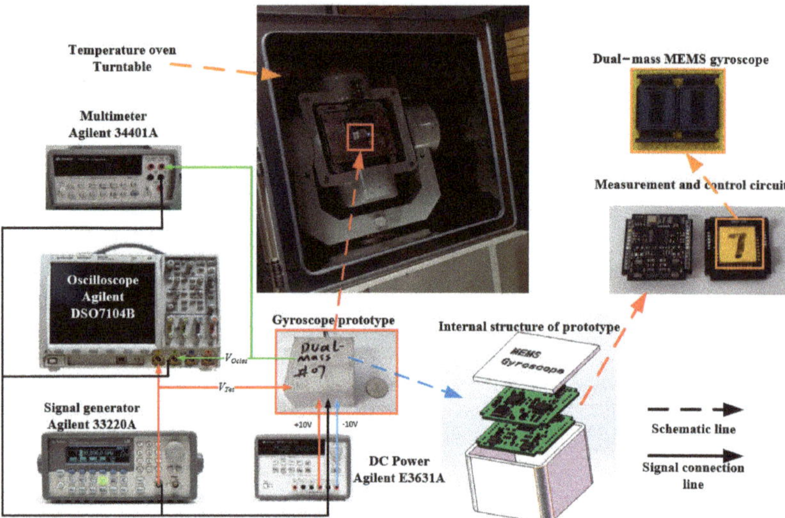

Figure 11. MEMS gyroscope porotype and temperature experiment platform.

In our experiment, we utilized a range of equipment to test the feasibility and efficacy of the MEMS gyroscope fault diagnosis platform. This equipment included an Agilent 33220A signal generator, which was capable of generating the test voltage, V_{Tes}. Additionally, we employed an Agilent 34401A multimeter and an Agilent DSO7104B oscilloscope to measure and observe signal amplitude and phase. For the power supply, we utilized an Agilent E3631A unit, which provided ± 10 V DC voltage and GND. Finally, to create a full range of temperature conditions for measuring the actual bandwidth of the gyroscope, we utilized both a temperature box and a turntable.

The experimental process proceeded as follows: The gyroscope was first turned on and allowed to run for one hour at room temperature. Subsequently, the oven temperature was gradually increased until it reached 60 °C. Then, to ensure a continuous and stable internal temperature in the gyroscope, the temperature should be maintained for one hour for every 10° drop from 60 °C to −40 °C and finally stand for one hour, the test was finished [35–37]. The above steps were repeated three times to obtain three groups of experimental results, as shown in Figure 12. For the three groups of drift signals, VMD−SE is used for feature extraction, and the ELM method is used for fault identification. The results of our identification process are presented in Figure 12. It is evident from the figure that the recognition rate based on real data is exceptionally high, reaching 100%. All of the data were successfully recognized, thus providing strong evidence for the reliability and accuracy of the proposed dualmass MEMS gyroscope fault diagnosis platform.

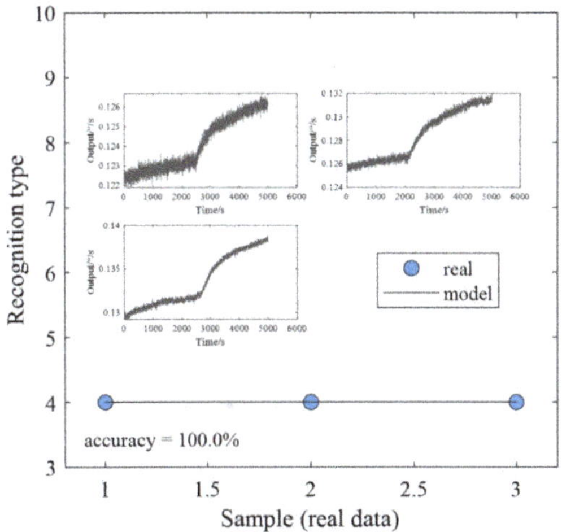

Figure 12. Identification results of real data.

6. Conclusions

Due to the inconvenience of gyroscope fabrication costs and fault dataset acquisition, in this study, a new Dualmass MEMS gyroscope fault diagnosis platform was designed. The platform integrates a dualmass MEMS gyroscope structural model and measurement control system, data feature extraction and classification prediction algorithm, and real data feedback verification. The accuracy rate reached up to 92.86% upon simulation training and test set verification, and all the effects of the platform were correctly identified through temperature experiments to obtain real data, indicating that the proposed fault diagnosis platform can accurately and effectively identify fault types.

Author Contributions: Conceptualization, M.Z.; Methodology, Z.W., J.X. and W.W.; Software, T.M., W.Z. and M.Z.; Investigation, H.C.; Resources, H.C.; Writing—original draft, R.C. and L.C.; Writing—review & editing, M.Z. All authors have read and agreed to the published version of the manuscript.

Funding: This work is supported by the National Key Research and Development Program of China (No. 2022YFB3205000), the National Natural Science Foundation of China, the NSAF (Grant No. U2230206), the Technology Field Fund of Basic Strengthening Plan of China (2020JCJQJJ409, 2021-JCJQ-JJ-0315), and the Pre-Research Field Foundation of Equipment Development Department of China (No. 61405170104 and No. 80917010501). The research is also supported by the Fundamental Research Program of Shanxi Province (20210302123020 and 20210302123062), the Shanxi Province Key Laboratory of Quantum Sensing and Precision Measurement (201905D121001), the Key Research and Development (R&D) Projects of Shanxi Province (202003D111004), the Beijing Key Laboratory of High Dynamic Navigation Technology Open Founding (HDN2021102), and the Fund for Shanxi "1331Project" Key Subjects Construction.

Data Availability Statement: Not applicable.

Conflicts of Interest: The authors declare no conflict of interest.

References

1. Shi, Y.; Wen, X.; Zhao, Y.; Zhao, R.; Cao, H.; Liu, J. Investigation and experiment of high shock packaging technology for High-G MEMS accelerometer. *IEEE Sens. J.* **2020**, *20*, 9029–9037. [CrossRef]
2. Farahani, H.V.; Rahimi, A. Data-Driven Fault Diagnosis for Satellite Control Moment Gyro Assembly with Multiple In-Phase Faults. *Electronics* **2021**, *10*, 1537. [CrossRef]
3. Benjemaa, R.; Elhsoumi, A.; Naoui, S.B.H.A. Active Fault Tolerant Control for uncertain neutral time delay system. In Proceedings of the 2020 17th International Multi-Conference on Systems, Signals & Devices (SSD), Monastir, Tunisia, 20–23 July 2020; pp. 285–289.

4. Beard, R.V. Failure Accommodation in Linear Systems through Self-Reorganization. Ph.D. Thesis, Rept, MVT-71-1, Man Vehicle Laboratory, Cambridge, MA, USA, 1971.

5. Luo, J.; Wen, G.; Lei, Z.; Su, Y.; Chen, X. Weak signal enhancement for rolling bearing fault diagnosis based on adaptive optimized VMD and SR under strong noise background. *Meas. Sci. Technol.* **2023**, *34*, 064001. [CrossRef]

6. Liu, W.; Li, S.; Chen, M.; Fang, Y.; Cha, L.; Wang, Z. Fault Diagnosis for Attitude Sensors Based on Analytical Redundancy and Wavelet Transform. In Proceedings of the Chinese Automation Congress, Shanghai, China, 6–8 November 2020.

7. Niu, R.; Liu, W.; Wang, B.; Li, L.; Wang, Z. Fault Diagnosis for Attitude Sensors based on Analytical Redundancy and EMD. In Proceedings of the 2019 Chinese Control and Decision Conference (CCDC), Nanchang, China, 3–5 June 2019.

8. Farahani, H.V.; Rahimi, A. Fault Diagnosis of Control Moment Gyroscope Using Optimized Support Vector Machine. In Proceedings of the 2020 IEEE International Conference on Systems, Man, and Cybernetics (SMC), Toronto, ON, Canada, 11–14 October 2020.

9. Liu, Y.; Wu, T.; Chen, J.; Fan, S.; Liu, X.; Gong, Y. Research on Local Mean Decomposition and Extreme Learning Machine based Circuit Breaker Fault Diagnosis Method. In Proceedings of the 2020 Asia Energy and Electrical Engineering Symposium (AEEES), Chengdu, China, 28–31 May 2020.

10. Zhang, W.; Zhang, D.; Zhang, P.; Han, L. A New Fusion Fault Diagnosis Method for Fiber Optic Gyroscopes. *Sensors* **2022**, *22*, 2877. [CrossRef] [PubMed]

11. Zhao, H.; Liu, M.; Sun, Y.; Chen, Z.; Duan, G.; Cao, X. Fault diagnosis of control moment gyroscope based on a new CNN scheme using attention enhanced convolutional block. *Sci. China Engl. Version Tech. Sci.* **2022**, *65*, 12. [CrossRef]

12. Cao, H. Design, Fabrication, and Experiment of a Decoupled Multi-Frame Vibration MEMS Gyroscope. *IEEE Sens. J.* **2021**, *21*, 19815–19824. [CrossRef]

13. Cao, H.; Li, H.; Shao, X.; Liu, Z.; Kou, Z.; Shan, Y.; Shi, Y.; Shen, C.; Liu, J. Sensing Mode Coupling Analysis for Dualmass MEMS Gyroscope and Bandwidth Expansion within Wide-Temperature Range. *Mech. Syst. Signal Process.* **2018**, *98*, 448–464. [CrossRef]

14. Cao, H.; Zhang, Y.; Han, Z.; Shao, X.; Gao, J.; Huang, K.; Shi, Y.; Tang, J.; Shen, C.; Liu, J. Pole-Zero Temperature Compensation Circuit Design and Experiment for Dualmass MEMS Gyroscope Bandwidth Expansion. *IEEE/ASME Trans. Mechatron.* **2019**, *24*, 677–688. [CrossRef]

15. Cao, H.; Li, H.; Liu, J.; Shi, Y.; Tang, J.; Shen, C. An improved interface and noise analysis of a turning fork microgyroscope structure. *Mech. Syst. Signal Process.* **2016**, *70–71*, 1209–1220. [CrossRef]

16. Cao, H.; Liu, Y.; Zhang, Y.; Shao, X.; Gao, J.; Huang, K.; Shi, Y.; Tang, J.; Shen, C.; Liu, J. Design and Experiment of Dualmass MEMS Gyroscope Sense Closed System Based on Bipole Compensation Method. *IEEE Access* **2019**, *7*, 49111–49124. [CrossRef]

17. Cao, H.; Xue, R.; Cai, Q.; Gao, J.; Zhao, R.; Shi, Y.; Huang, K.; Shao, X.; Shen, C. Design and Experiment for Dualmass MEMS Gyroscope Sensing closed-loop System. *IEEE Access* **2020**, *8*, 48074–48087. [CrossRef]

18. Cao, H.; Liu, Y.; Kou, Z.; Zhang, Y.; Shao, X.; Gao, J.; Huang, K.; Shi, Y.; Tang, J.; Shen, C.; et al. Design, Fabrication and Experiment of Double U-beam MEMS Vibration Ring Gyroscope. *Micrimachines* **2019**, *10*, 186. [CrossRef] [PubMed]

19. Cao, H.; Li, H.; Kou, Z.; Shi, Y.; Tang, J.; Ma, Z.; Shen, C.; Liu, J. Optimization and experimentation of dualmass MEMS gyroscope quadrature error correction methods. *Sensors* **2016**, *16*, 71. [CrossRef] [PubMed]

20. Shi, Y.; Wang, Y.; Feng, H.; Zhao, R.; Cao, H.; Liu, J. Design, fabrication and test of a low range capacitive accelerometer with anti-overload characteristics. *IEEE Access* **2020**, *8*, 26085–26093. [CrossRef]

21. Cao, H.; Zhang, Z.; Zheng, Y.; Guo, H.; Zhao, R.; Shi, Y.; Chou, X.; García, G.-P.A. A New Joint Denoising Algorithm for High-G Calibration of MEMS Accelerometer Based on VMD-PE-Wavelet Threshold. *Shock Vib.* **2021**, *2021*, 8855878. [CrossRef]

22. Chang, L.; Cao, H.; Shen, C. Dualmass MEMS Gyroscope Parallel Denoising and Temperature Compensation Processing Based on WLMP and CS-SVR. *Micromachines* **2020**, *11*, 586. [CrossRef]

23. Shi, J.; Zhou, J.; Feng, J.; Chen, H. Network traffic prediction model based on improved VMD and PSO-ELM. *Int. J. Commun. Syst.* **2023**, *36*, e5448. [CrossRef]

24. Wang, Z.; Yao, L.; Chen, G.; Ding, J. Modified multiscale weighted permutation entropy and optimized support vector machine method for rolling bearing fault diagnosis with complex signals. *ISA Trans.* **2021**, *114*, 470–484. [CrossRef]

25. Cao, H.; Liu, Y.; Liu, L.; Wang, X. Humidity Drift Modeling and Compensation of MEMS Gyroscope Based on IAWTD CSVM EEMD Algorithms. *IEEE Access* **2021**, *9*, 95686–95701. [CrossRef]

26. Ni, N.; Dong, S. Numerical Computation of Partial Differential Equations by Hidden-Layer Concatenated Extreme Learning Machine. *J. Sci. Comput.* **2022**, *95*, 35. [CrossRef]

27. Yu, R.; Li, X.; Tao, M.; Ke, Z. Fault Diagnosis of Feedwater Pump in Nuclear Power Plants Using Parameter-Optimized Support Vector Machine. In Proceedings of the 2016 24th International Conference on Nuclear Engineering, Charlotte, NC, USA, 26–30 June 2016.

28. Kalita, K. Analyzing Physics-Inspired Metaheuristic Algorithms in Feature Selection with K-Nearest-Neighbor. *Appl. Sci.* **2023**, *13*, 906.

29. Redivo, E.; Viroli, C.; Farcomeni, A. Quantile distribution functions and their use for classification with application to nave bayes classifiers. *Stat. Comput.* **2023**, *33*, 55. [CrossRef]

30. Alrashidi, M. Estimation of Weibull Distribution Parameters for Wind Speed Characteristics Using Neural Network Algorithm. *CMC Comput. Mater. Con.* **2023**, *75*, 1073–1088. [CrossRef]

31. Hassanien, A.E. Classification and Feature Selection of Breast Cancer Data Based on Decision Tree Algorithm. *Stud. Inform. Control.* **2003**, *12*, 33–40.
32. Cao, H.; Wei, W.; Liu, L.; Ma, T.; Zhang, Z.; Zhang, W.; Shen, C.; Duan, X. A Temperature Compensation Approach for Dualmass MEMS Gyroscope Based on PE LCD and ANFIS. *IEEE Access* **2021**, *9*, 95180–95193. [CrossRef]
33. Cao, H.; Li, H.; Sheng, X.; Wang, S.; Yang, B.; Huang, L. A novel temperature compensation method for MEMS gyroscope's oriented on periphery circuit. *Int. J. Adv. Robot. Syst.* **2013**, *10*, 327. [CrossRef]
34. Cao, H.; Li, H. Investigation of a vacuum packaged MEMS gyroscope architecture's temperature robustness. *Int. J. Appl. Electromagn. Mech.* **2013**, *41*, 495–506. [CrossRef]
35. Cao, H.; Zhang, Y.; Shen, C.; Liu, Y.; Wang, X. Temperature Energy Influence Compensation for MEMS Vibration Gyroscope Based on RBF NN-GA-KF Method. *Shock. Vib.* **2018**, *2018*, 2830686. [CrossRef]
36. Ma, T.; Li, Z.; Cao, H.; Shen, C.; Wang, Z. A Parallel Denoising Model for Dualmass MEMS Gyroscope Based on PE-ITD and SA-ELM. *IEEE Access* **2019**, *7*, 113901–113915.
37. Ma, T.; Cao, H.; Shen, C. A Temperature Error Parallel Processing Model for MEMS Gyroscope Based on a Novel Fusion Algorithm. *Electronics* **2020**, *9*, 499. [CrossRef]

micromachines

MDPI

Article

Temperature Drift Compensation for Four-Mass Vibration MEMS Gyroscope Based on EMD and Hybrid Filtering Fusion Method

Zhong Li [1,2], **Yuchen Cui** [3], **Yikuan Gu** [1,2], **Guodong Wang** [4], **Jian Yang** [1,2], **Kai Chen** [5] **and Huiliang Cao** [6,*]

[1] Shanxi Software Engineering Technology Research Center, Taiyuan 030051, China
[2] School of Software, North University of China, Taiyuan 030051, China
[3] School of Instrument and Electronics, North University of China, Taiyuan 030051, China
[4] Beijing Institute of Aerospace Control Devices, Beijing 100039, China
[5] School of Automation Engineering, University of Electronic Science and Technology of China, Chengdu 611731, China
[6] Key Laboratory of Instrumentation Science & Dynamic Measurement, Ministry of Education, North University of China, Taiyuan 030051, China
* Correspondence: caohuiliang@nuc.edu.cn

Abstract: This paper presents an improved empirical modal decomposition (EMD) method to eliminate the influence of the external environment, accurately compensate for the temperature drift of MEMS gyroscopes, and improve their accuracy. This new fusion algorithm combines empirical mode decomposition (EMD), a radial basis function neural network (RBF NN), a genetic algorithm (GA), and a Kalman filter (KF). First, the working principle of a newly designed four-mass vibration MEMS gyroscope (FMVMG) structure is given. The specific dimensions of the FMVMG are also given through calculation. Second, finite element analysis is carried out. The simulation results show that the FMVMG has two working modes: a driving mode and a sensing mode. The resonant frequency of the driving mode is 30,740 Hz, and the resonant frequency of the sensing mode is 30,886 Hz. The frequency separation between the two modes is 146 Hz. Moreover, a temperature experiment is performed to record the output value of the FMVMG, and the proposed fusion algorithm is used to analyse and optimise the output value of the FMVMG. The processing results show that the EMD-based RBF NN+GA+KF fusion algorithm can compensate for the temperature drift of the FMVMG effectively. The final result indicates that the random walk is reduced from $99.608°/h/Hz^{1/2}$ to $0.967814°/h/Hz^{1/2}$, and the bias stability is decreased from $34.66°/h$ to $3.589°/h$. This result shows that the algorithm has strong adaptability to temperature changes, and its performance is significantly better than that of an RBF NN and EMD in compensating for the FMVMG temperature drift and eliminating the effect of temperature changes.

Keywords: four-mass vibration MEMS gyroscope; theoretical simulation; temperature drift compensation; empirical mode decomposition; radial basis function neural network; genetic algorithm; Kalman filter

Citation: Li, Z.; Cui, Y.; Gu, Y.; Wang, G.; Yang, J.; Chen, K.; Cao, H. Temperature Drift Compensation for Four-Mass Vibration MEMS Gyroscope Based on EMD and Hybrid Filtering Fusion Method. *Micromachines* **2023**, *14*, 971. https://doi.org/10.3390/mi14050971

Academic Editor: Nam-Trung Nguyen

Received: 19 April 2023
Revised: 26 April 2023
Accepted: 27 April 2023
Published: 28 April 2023

1. Introduction

MEMS (micro-electro-mechanical systems) gyroscopes are widely used in both civil and military fields. They are mainly used in fields such as automobile navigation, inertial navigation, attitude determination, robotics, micro-signal detection, aerospace and spatial positioning, and application electronics [1–6]. However, temperature variation can seriously reduce the precision of the MEMS gyroscope when used in a changing environment. Therefore, it is necessary to study how to eliminate the influence of temperature drift on MEMS gyroscopes. In general, there are two main methods for the temperature compensation of MEMS gyroscopes: hardware compensation methods and software compensation methods.

The hardware method is mainly used to optimise the structure of the MEMS gyroscope. For example, Yu proposed to use a polarising resonator to improve the thermal stability of a resonator fibre optic gyro [7]. Gao proposed that by calculating the optimal layer number of the fibre coil, the temperature drift of the interferometric fibre optic gyroscope can be effectively reduced [8]. Wang proposed that the use of a symmetrical push-pull structure can eliminate the effects of temperature drift [9]. Yan proposed to use a single-polarisation fibre, and the resonant fibre optic gyro was found to exhibit good stability over a wide temperature range [10]. Ling proposed a novel winding idea of dicyclic fibre coils for suppressing the thermal-induced bias drift error of interferometric fibre optic gyroscopes [11]. Qian proposed a hybrid resonator consisting of a polymer-based long-range surface plasmon polariton (LRSPP) waveguide coupler and a silica fibre to reduce the effects of temperature noise on the resonant fibre optic gyroscope [12]. Zhang proposed an improved QAD fibre coil, which helped to reduce the thermal-induced drift error of a fibre optic gyroscope and improved its precision [13]. Zhang proposed a special resonator structure that can generate oriented thermal stress to reduce the temperature drift of the vibrating axis of the resonator [14]. To some extent, hardware compensation methods have the disadvantage of increasing the size and power consumption and cost of MEMS gyroscopes, so these methods are not suitable for some applications that require low cost and low power.

The second solution is a software compensation method that takes as its input the relationship between the temperature of the MEMS gyroscope and the output signal and then establishes a temperature drift model to compensate for the effects of temperature drift. For example, Fontanella proposed a new method for accurate thermal compensation of the inertial device gyro bias using an augmented state Kalman filter [15]. Cao proposed a fusion algorithm based on an RBF NN-GA-KF, which effectively reduced the effect of the temperature drift on the MEMS vibration gyroscope [16]. Baxpameeb proposed an improved method for real-time thermal drift compensation of fibre optic gyros, through which the fibre gyro temperature drift was effectively compensated [17]. Narasimhappa proposed an ARMA model based on an adaptive unscented fading Kalman filter to reduce the drift of a fibre-optic gyroscope [18]. Zhang proposed a novel wavelet threshold denoising method, which effectively suppresses the environmental drift of fibre optic gyroscopes [19]. Song proposed a new modelling and compensation method combining an artificial fish swarm algorithm (AFSA) and a back-propagation (BP) neural network to improve the output precision of fibre optic gyroscopes [20]. Han proposed a simplified model to compensate for the output of ring laser gyros, which effectively improved the output accuracy of the gyroscopes [21]. Ahtohoba used factors such as the temperature change rate of the environment; an error model of the fibre optic gyroscope was established, and the output precision of the gyro was ultimately improved [22]. Zha established an IUKF neural network model for temperature drift compensation of the fibre optic gyro [23]. Prikhodko proposed to compensate high-Q MEMS gyroscopes by using temperature self-sensing technology, and the compensated gyroscope has high precision [24]. Chen proposed a fibre optic gyro adaptive positive linear prediction denoising algorithm based on the ambient temperature rate of change and a temperature drift modelling compensation concept, which was used to correct errors caused by drastic changes in the ambient temperature [25]. Wang proposed to establish an HRG temperature compensation model based on the natural frequency of the resonator and then used this model to compensate for the temperature drift of the gyroscope; this method can be used over a wide temperature range [26]. Cai established a parallel processing model based on multi-objective particle swarm optimisation based on a variational modal decomposition-time-frequency peak filter (MOVMD-TFPF) and Beetle antennae search algorithm-Elman neural network (BAS-Elman NN) to eliminate the noise and temperature drift in a Micro-Electro-Mechanical Systems (MEMS) gyroscope's output signal [27]. Zhou proposed the improved VMD and TFPF hybrid denoising algorithm, which combines variational modal decomposition (VMD) and time-frequency peak filtering (TFPF), which has a smaller signal distortion and stronger denoising ability [28]. In recent years, the academic circle has shown great interest in the research of a four-mass gyroscope. Zhou proposed a novel Center Support Quadruple Mass Gyroscope, which combines the advantages of tuning fork gyros and micro hemisphere resonant gyros and is expected to achieve performance breakthroughs of flat structures [29]. Trusov proposed a new tuning fork architecture addressing the limitations of the conventional designs [30]. The software method has the advantages of a simple structure, low cost, and convenient implementation, and it is convenient for improving the precision of the MEMS gyroscope.

In this paper, an improved RBF NN filtering algorithm based on EMD is proposed, and this method has been successfully applied to eliminate the temperature drift of the MEMS gyro. Different from the other MEMS gyroscope temperature compensation algorithms, this paper proposes a new fusion algorithm (EMD-based RBF NN+GA+KF fusion algorithm) to eliminate the influence of temperature drift on MEMS gyroscopes and make the MEMS gyroscope more accurate.

The structure of the MEMS gyroscope is introduced in Section 2; the fusion algorithm is shown in Section 3; Section 4 describes the temperature experiment; and Section 5 gives the results of the data processing and a comparison of the various algorithms. The conclusion of this paper is given in Section 6.

2. Structure of the Four-Mass Vibration MEMS Gyroscope

The original signal is generated by a four-mass vibration MEMS gyroscope (FMVMG). The schematic diagram of the FMVMG is shown in Figure 1a. The FMVMG is composed of support anchor points, quadruple mass blocks, a drive comb, a drive sense comb, a sense comb, a force rebalances comb, and multiple folding beams. This gyroscope has two operating modes: a drive mode and a sense mode. Both modes can be considered as a "mass-spring-damped" second-order vibration system. The basic mechanical equivalent model is shown in Figure 1b.

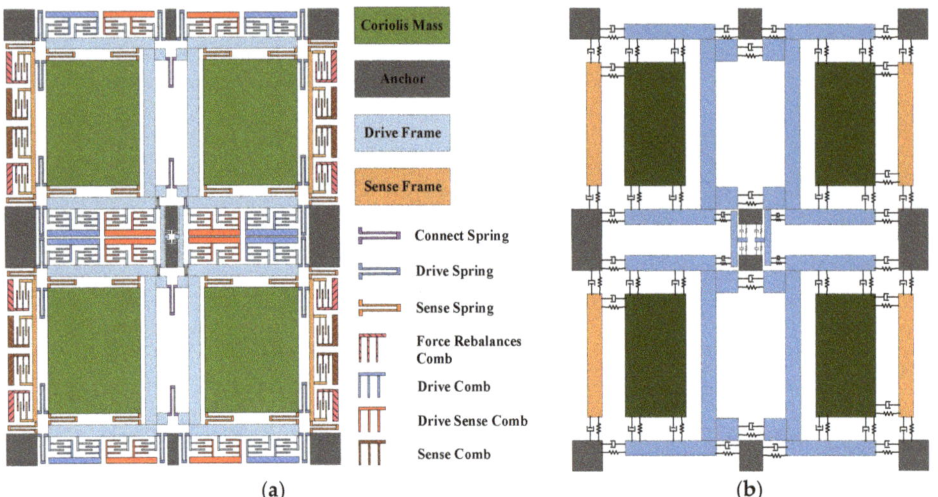

Figure 1. (**a**) Structure of FMVMG, (**b**) The lumped model of FMVMG.

Firstly, after the preliminary design and optimisation, the geometric shape of the FMVMG was determined, and the specific parameters are shown in Table 1. Then, the modal component in ANSYS software was used to analyse the modal of the FMVMG, and the modal analysis mainly included the natural frequency, modal shape, and vibration stability of the MEMS gyroscope structure. Finally, the driving and sensing mode shapes of the FMVMG are shown in Figure 2.

Table 1. MEMS gyroscope structure mechanical values.

Parameter	Value
Elastic modulus (E)	169 GPa
Poisson's ratio (μ)	0.27
Density (ρ)	2328.3 kg/m^3
Prototype length	3015 µm
Prototype width	2331 µm
Prototype height	60 µm

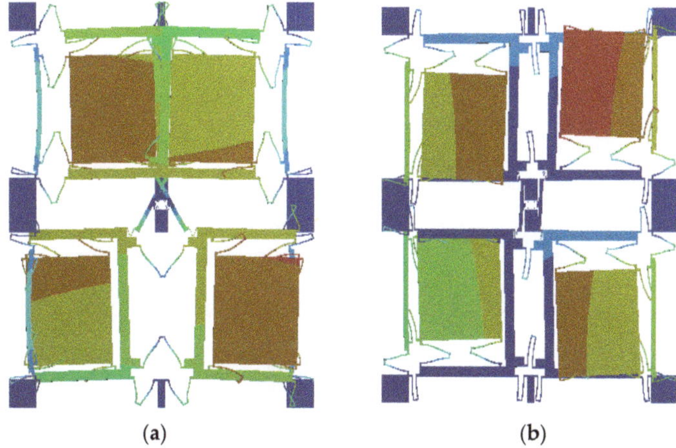

(a)   (b)

Figure 2. (a) Driving mode, (b) Sensing mode.

As shown in Figure 2, the vibration mode of the driving mode is the motion of the Coriolis mass block on the X-axis, while the vibration mode of the sensing mode is the motion of the Coriolis mass block on the Y-axis. In both the driving and sensing modes, the motion of the adjacent mass blocks is reversed. Among them, the resonant frequency of the driving mode is 30,740 Hz, and the resonant frequency of the sensing mode is 30,886 Hz due to the highest mechanical sensitivity when the natural frequencies of the driving and sensing modes of the MEMS gyroscope are equal, but the bandwidth is small. The frequency difference of this structure is 146Hz, which not only ensures the mechanical sensitivity of the gyroscope but also gives it a certain bandwidth.

The gyroscope control and detection system are shown in Figure 3. In the drive loop, the drive frame displacement, $x(t)$, is detected by the drive sensing combs and picked up by the differential amplifier ①. Then, the signal phase is delayed by 90° (through ②) to satisfy the phase requirement of the AC drive signal, $V_{dac}Sin(\omega_d t)$. After that, $V_{dac}Sin(\omega_d t)$ is processed by a full-wave rectifier ③ and a low pass filter ④. Afterwards, V_{dac} is compared (in ⑤) with the reference voltage, V_{ref} ⑥. Next, the drive PI controller ⑦ generates the control signal, which is modulated by $V_{dac}Sin(\omega_d t)$, and then the signal is superposed (through ⑩) by V_{DC} ⑨ to stimulation drive mode.

The sensing system employs an open loop, which utilises the same interface as the drive circuit. Firstly, the left and right masses' sensing signals are detected separately with the differential detection amplifier ⑪. In addition, the output signals are processed by a second differential amplifier ⑫ to generate a signal, V_{stotal}. Then, V_{stotal} is demodulated by signal $V_{dac}Sin(\omega_d t)$ (in ⑬). After that, the demodulated signal, V_{dem}, passes through the low pass filter ⑭ so the sensing mode's movement signal, V_{Oopen}, can be obtained.

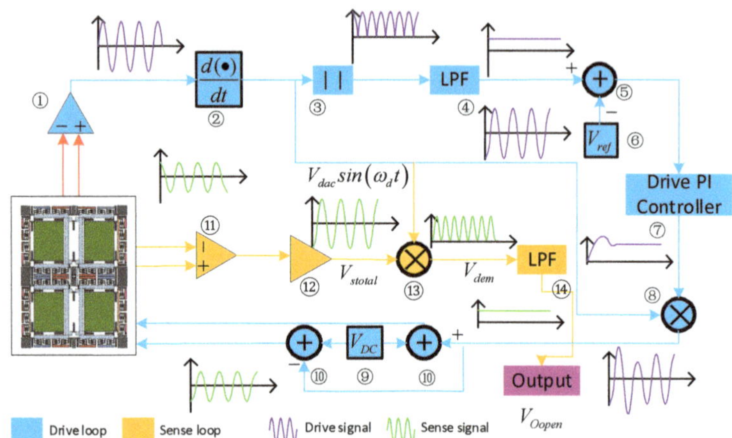

Figure 3. FMVMG monitoring system.

The monitoring system consists of three PCBs. PCBI contains the connection circuit and is connected to the structure chip, and PCB II is the drive circuit. PCB I and PCB III are nested on the top and back sides, respectively. PCB III contains the detection loop, and the output signal is from PCB III. The prototype of the FMVMG is shown in Figure 4.

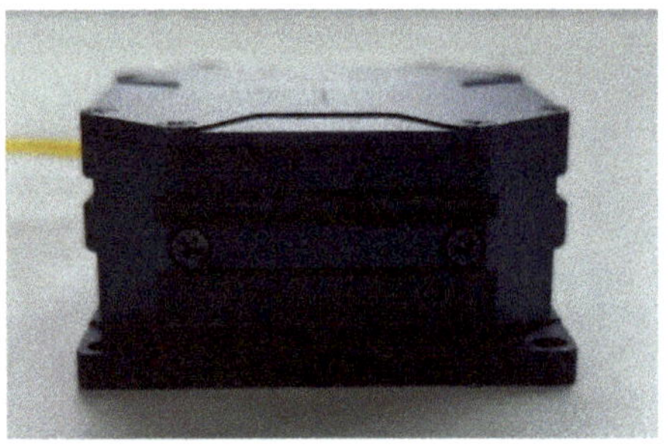

Figure 4. FMVMG prototype.

3. Model and Algorithm

3.1. SE-EMD Algorithm

Empirical mode decomposition is a new adaptive signal time-frequency processing method that is especially suitable for the analysis and processing of nonlinear non-stationary signals. In essence, this method is also a process of smoothing non-stationary signals by generating a series of data sequences, and their feature scales are different. Each of these data sequences is called an intrinsic mode function (IMF). The specific steps of the EMD are as follows:

(1) Take out all the local extremum points of the signal and use the cubic spline curve to connect all the local maxima to form the upper envelope. Similarly, the minimum values are also connected to form the lower envelope. In addition, find the average value of the upper and lower envelopes, denoted as m_1, and h_1 can be obtained as follows:

$$x(t) - m_1 = h_1 \tag{1}$$

where $x(t)$ is the original data, and if h_1 meets the IMF component conditions, then h_1 can be seen as the first IMF component of $x(t)$.

(2) If h_1 is not the IMF, set h_1 as the original data, continue with step 1, obtain the average value m_{11}, and evaluate $h_{11} = h_1 - m_{11}$; if h_{11} is still not an IMF component, repeat the loop k times and obtain $h_{1(k-1)} - m_{1k} = h_{1k}$ until h_{1k} meets the conditions of the IMF. Note that $c_1 = h_{1k}$. Then, c_1 is the first of the signals $x(t)$ that meets the IMF condition.

(3) Separate c_1 from $x(t)$ and obtain Equation (2):

$$r_1 = x(t) - c_1 \tag{2}$$

(4) Substitute the value solved by Equation (2) into Step 1 and Step 2, obtain the second IMF-compliant component, and repeat this step n times to obtain the nth IMF-compliant components of the signal, namely:

$$\begin{cases} r_1 - c_2 = r_2 \\ \quad \cdots \\ r_{n-1} - c_n = r_n \end{cases} \tag{3}$$

(5) If the resulting r_n is a monotonic function and no additional IMF-compliant components can be derived from it, the steps are not repeated. The final result of EMD decomposition is shown in Equation (4):

$$x(t) = \sum_{i=1}^{n} c_i + r_n \tag{4}$$

where $c_1, c_2, c_3, \ldots, c_n$ are the individual IMF components, and r_n represents the average trend of the signal.

After EMD decomposition, the output signal of the gyroscope is decomposed into many IMF components. To simplify the analysis, the IMFs are divided into three component types: a noise component, a mixed component, and a drift component. The sample entropy (SE) is introduced here to realise this work. The sample entropy is a measure of the complexity of a time series. The larger the value of SE is, the more complex and irregular the data. SE can be defined as follows:

$$SE(m,r) = \lim_{N \to \infty} \left\{ -ln[\frac{A^m r}{B^m r}] \right\} \tag{5}$$

where m is the embedding dimension, referred to as the length of the data to be compared, and r is the tolerance value.

3.2. The Algorithm of RBF NN+GA+KF

The radial basis function (RBF) is a non-negative nonlinear function characterised by a local distribution of the centre point and a symmetric attenuation of the radial centre point. The smaller the distance between the input point and the centre of the radial basis function, the stronger the output signal.

The basis function of the RBF NN is shown in Equation (6):

$$\alpha_j(x) = \psi_j(\|x - c_j\|/\sigma_j) \tag{6}$$

where x is an n-dimensional input vector, c_j is the centre of the jth hidden node, σ_j is the breadth of the hidden layer neuron, $\|.\|$ is the Euclidean distance, and ψ_j has a maximum value at c_j, which is a radially symmetric function. When $\|x - c_j\|$ increases, ψ_j decreases quickly.

To establish an RBF neural network model, an RBF is introduced into the hidden layer of the neural network. The RBF NN structure is shown in Figure 5, which is similar to the other forward neural networks. The RBF neural network has three layers: the input layer, the hidden layer, and the output layer. The input layer consists of signal source nodes. The number of elements in the hidden layer is determined according to the specific problem, and the hidden layer centre transform function is a radial basis function. The output layer is the response to the input mode.

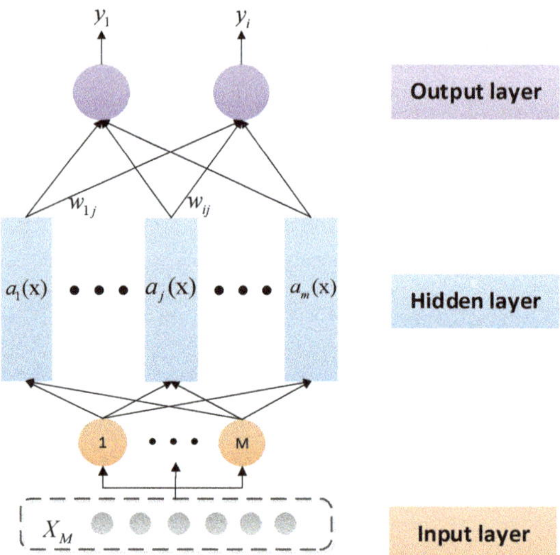

Figure 5. Model of a radial basis function neural network.

Here, the Gaussian basis function is chosen as the transfer function of the RBF neurons, which is shown in Equation (7):

$$a_j(x) = \psi_j(\|x - c_j\|/\sigma_j) = e^{-\frac{\|x-c_j\|^2}{2\sigma_j^2}} \tag{7}$$

The output of the RBF NN is shown in Equation (8):

$$y_i = \sum_{j=1}^{m} w_{ij}\alpha_j(x), i = 1, 2, \cdots, r \tag{8}$$

where W_{ij} is the connection weight between the *ith* hidden layer node and the *jth* output layer node. The learning algorithm is shown in Equation (9):

$$w_{ij}(l+1) = w_{ij}(l) + \beta[y_i^d - y_i(l)]\alpha_j(x)/\alpha^T(x)\alpha(x) \tag{9}$$

where β is the learning rate. When β falls within the range of 0 to 2, the algorithm offers good convergence performance.

During the training process of the RBF NN, the local amplification of the RBF neural network occurs. If x is far from c_j, the output of $\alpha_j(x)$ is also nearly 0 when passing through the linear nerve of the second layer. When x and c_j become very close to 0, the output of the layer is almost equal to 1, and the output value is almost equal to the weight value on the second layer when passing through the second layer

The RBF neural network has the characteristics of a simple structure, fast operation speed, and local function approximation. In general, the more hidden layer neuron nodes of the RBF neural network, the stronger the computing power and mapping ability are and the better the function is. The approximation ability can approach a complex function curve with arbitrary precision.

However, the performance of the RBF NN model depends mainly on the length of the training sample set and the expansion constant of the radial basis function (breadth σ_j). The length of the training sample set affects the performance of the network and the real-time performance of the training directly; if the expansion constant, σ_j, is too small, the network will over-fit, and, conversely, if it is too large, the network will not have the ability to fit. Therefore, genetic algorithms are introduced to solve this problem.

The genetic algorithm is an adaptive global probability search algorithm that simulates the inheritance and survival of creatures in their natural environment. The algorithm mainly includes five basic steps: coding, initial population generation, fitness setting function, genetic operator design, and operational parameter determination. The process of a GA is shown in Figure 6. In combining the GA with the RBF neural network, use the GA to solve the parameter selection problem of the RBF NN model. The specific algorithm is as follows: set the maximum algebra, G_{MAX}, of the genetics and combine each body in the population as RBF NN training parameters to find the corresponding output error. After training the algebra maximum, stop the calculation.

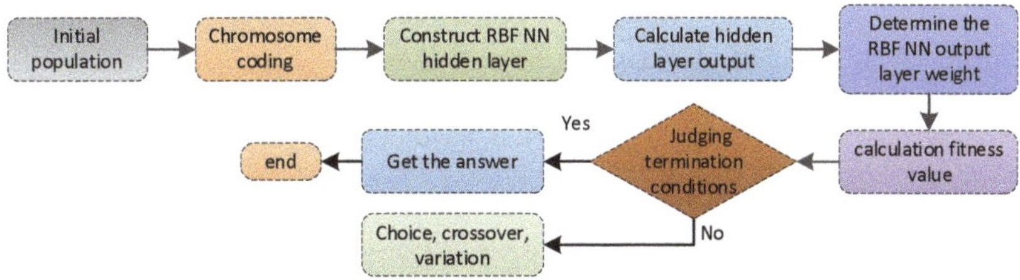

Figure 6. The process of the RBF NN based on a GA (RBF NN+GA).

The Kalman filter algorithm is an optimal estimation of the autoregressive data processing algorithm. It is widely used in the elimination of gyro noise because of its high estimation accuracy, fast convergence speed, strong adaptability, and simple calculation. Since the original data contain the influence of noise, the optimal choice of data processing can be regarded as a filtering process. The Kalman filtering process can be approximated as the following five equations:

$$\hat{X}_{k/k-1} = \Phi_{k,k-1}\hat{X}_{k-1} \tag{10}$$

$$\hat{X}_k = \hat{X}_{k/k-1} + K_k(Z_k - H_k\hat{X}_{k/k-1}) \tag{11}$$

$$K_k = P_{k/k-1}H_k^T(H_kP_{k/k-1}H_k^T + R)^{-1} \tag{12}$$

$$P_{k/k-1} = \Phi_{k,k-1}P_{k-1}\Phi_{k,k-1}^T + \Gamma_{k-1}Q\Gamma_{k-1}^T \tag{13}$$

$$P_k = (I - K_kH_k)P_{k/k-1} \tag{14}$$

where H_k is a measurement matrix, $P_{k/k-1}$ is a matrix of the prior estimate error covariance, P_k is a matrix of the estimation error variance, K_k is a filter gain matrix, R is the covariance of the measurement noise, and Q is the covariance of the process noise. According to the above equations, it can be found that the state estimation of $\hat{X}_k(k = 1, 2, \cdots)$ at time point k can be calculated by the measurement vector of Z_k at time point k if the initial value of $\hat{X}_0$ and P_0 are determined.

In this section, the RBF NN, GA, and KF are combined, and the RBF NN+GA+KF fusion algorithm is proposed. The algorithm has the advantages of online real-time use, training capability, and fast learning. The fusion algorithm of RBF NN+GA+KF is shown in Figure 7.

Figure 7. The process of RBF NN+GA+KF.

3.3. RBF NN+GA+KF Denoising Based on SE-EMD

After being decomposed by EMD, the original data are decomposed into multiple IMF components. Each IMF component is classified by using the sample entropy (SE). Generally, these IMF components can be divided into three groups: the noise component, the mixture component, and the drift component. The noise has the characteristics of high frequency and low drift. The high-frequency noise component of the IMFS can be removed directly. However, the mixed component contains both the useful signal and the equivalent noise signal, so the RBF NN+GA+KF is applied to the filtering of this component, which is beneficial for protecting the useful signal of the gyroscope and removing the noise signal effectively. The processing of the drift component can be divided into two steps. First, the output of the temperature sensor is decomposed by EMD to obtain the trend term of the temperature, and then the trend term of the temperature is used to compensate for the drift component of the gyroscope. Finally, the signal reconstruction of the MEMS gyroscope is performed by using the mixed component after denoising and the drift component after compensation by the trend term of the temperature. The specific flow chart is shown in Figure 8.

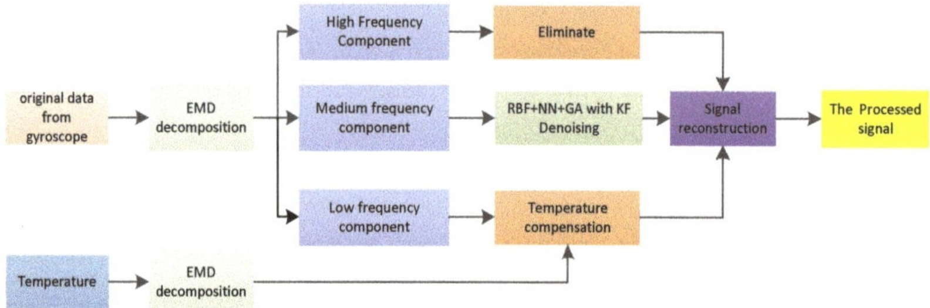

Figure 8. The process of RBF NN+GA+KF denoising based on SE-EMD.

4. Temperature Experiment Proposal

Through temperature experiments, the temperature characteristics of the MEMS gyroscope were tested. In the temperature experiment, the temperature-controlled oven can accurately provide a temperature range of $-40\,°\mathrm{C}$ to $+150\,°\mathrm{C}$, as shown in Figure 9. The gyroscope output is collected by dedicated data acquisition software. Power is provided by a Gwinstek GPS-4303C DC power supply. The real-time temperature in the gyroscope metal casing is obtained by a temperature sensor, whose value is synchronised with the gyroscope output. The indoor temperature is $25\,°\mathrm{C}$.

Firstly, the gyroscope is fixed to the static plane to avoid the influence of motion. Secondly, the output lines are connected to the Gwinstek DC power supply and the laptop. Then, the oven temperature is set at $-30\,°\mathrm{C}$ to $+60\,°\mathrm{C}$. Finally, the power is switched on, and the original output signal of the gyroscope is recorded. The data collection process is continuous, and the temperature experiment results are shown in Figure 10. It can be seen that the gyroscope output changes more obviously when the temperature changes. This phenomenon indicates that the temperature affects the precision of the gyroscope output seriously. Therefore, it is necessary to establish a temperature drift model of the MEMS gyroscope.

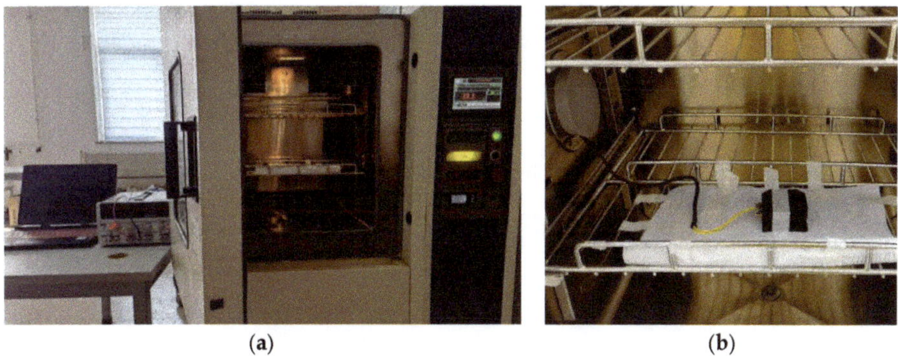

(a)   (b)

Figure 9. (**a**) Temperature experiment equipment. (**b**) Inside the oven.

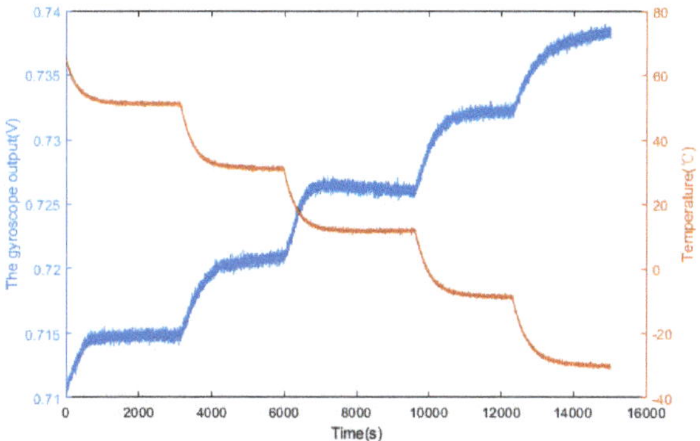

Figure 10. Temperature experiment results in a temperature-controlled oven.

5. Verification and Analysis

Figure 11 shows the results of the EMD processing of the gyroscope temperature drift error, which is decomposed into 7 IMF components and a residual component. These IMF components are characterised by non-stationary and complex components. For the convenience of processing, the sample entropy (SE) method is used to classify the IMF components of each layer, which can reduce the cumulative error when processing a single IMF reconstruction. Figure 12 shows the feature extraction based on the SE. It can be seen from Figure 13 that these IMF components can be roughly divided into three groups: IMF1, IMF2 as a group; IMF3, IMF4 as a group; and IMF5 to IMF7 as a group. These three groups are the three signal components mentioned above: the noise component ($S1$), the mixed component ($S2$), and the drift component ($S3$). The three signals are processed separately below.

According to the temperature error handling method proposed in Figure 8, first, the noise component ($S1$) is removed due to the high-frequency and low-drift characteristics of the noise, and the mixed component ($S2$) is denoised by the RBF NN+GA+KF algorithm. The denoised results are shown in Figure 14a. Second, after the temperature sensor output signal is decomposed by EMD, as shown in Figure 13, the temperature trend is obtained. Then, the temperature component is used to compensate for the drift component of the gyroscope output. Finally, the $S2$ and $S3$ component signals after processing are reconstructed, and the reconstruction results are shown in Figure 14b.

It can be seen from Figure 14 that the EMD improved algorithm proposed in this paper can effectively eliminate the error caused by temperature drift.

To compare the temperature compensation results quantitatively, the Allen analysis of variance was introduced. The Allen analysis is an analysis method that is based on the time domain. The main feature of the Allan variance method is that it can easily characterise and identify various error sources and their contribution to the overall noise statistics, and it also has the advantages of easy calculation and easy separation. Figure 15 shows the different compensation results for the Allan standard deviation curve, and Table 2 gives the quantitative results. It can be concluded from Table 2 that the proposed EMD-based RBF NN+GA+KF fusion method has the best performance, and the bias stability is increased from $34.66°/h$ to $3.589°/h$. The angular random walk is reduced from $99.608°/h/Hz^{1/2}$ to $0.967814°/h/Hz^{1/2}$.

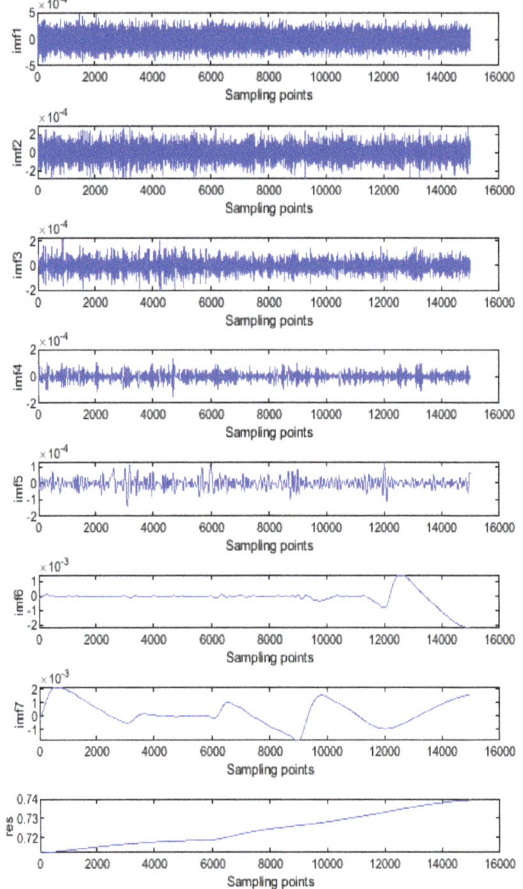

Figure 11. IMFs decomposed by EMD (gyroscope output).

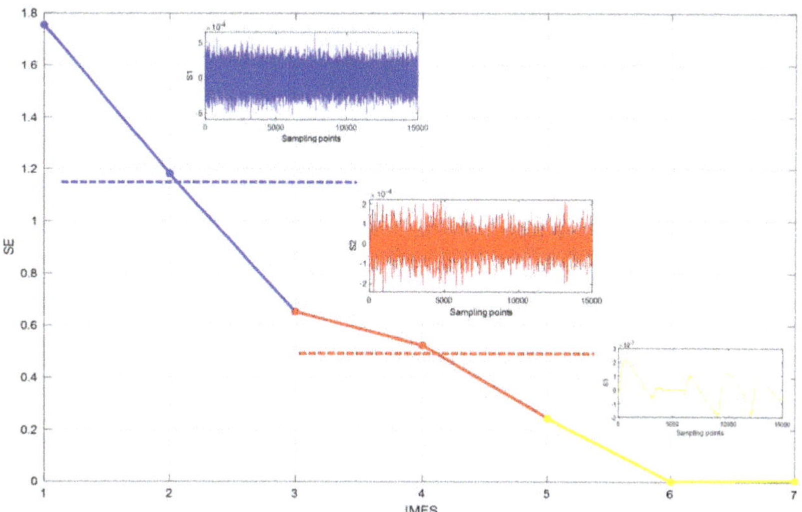

Figure 12. Feature component extraction based on the SE.

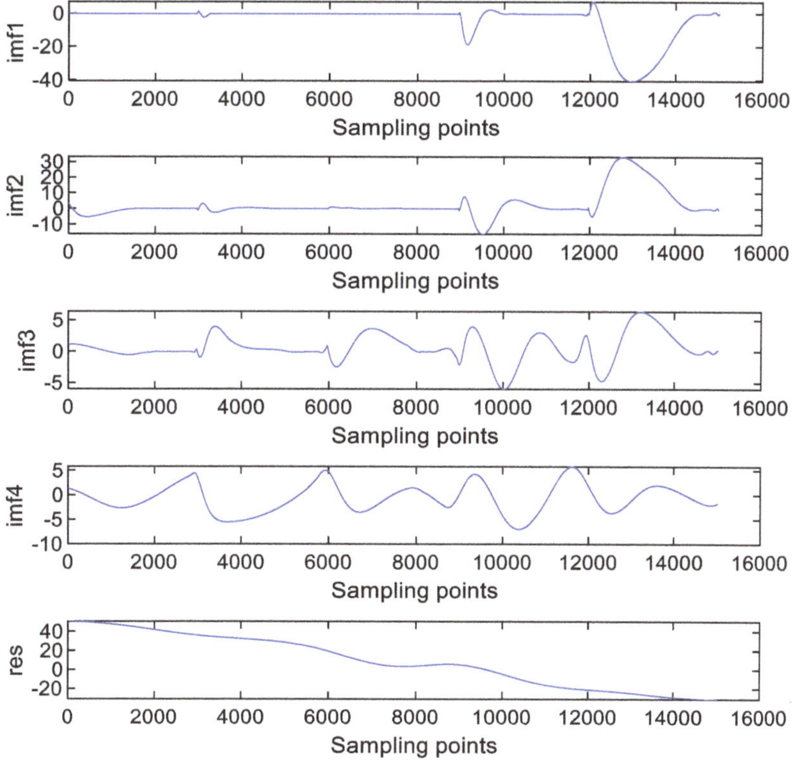

Figure 13. IMFs decomposed by EMD (temperature).

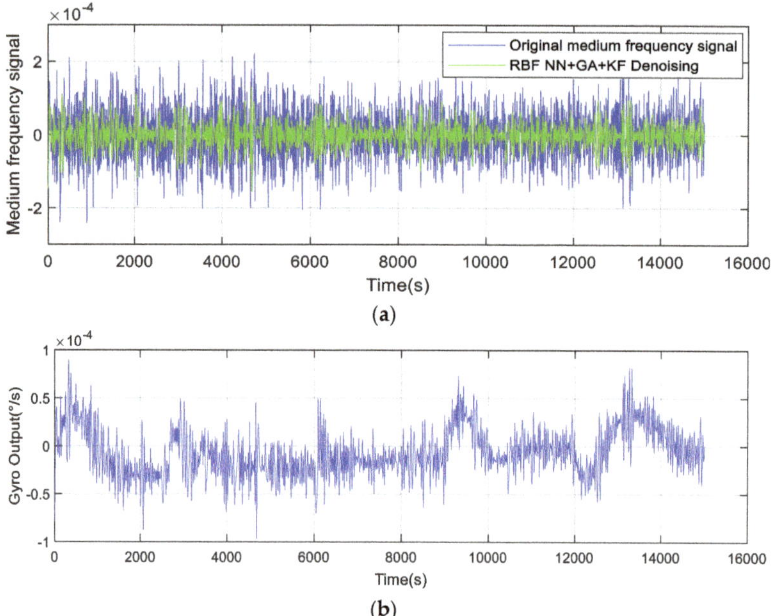

Figure 14. (**a**) Mixed component after denoising, (**b**) The final reconstructed signal.

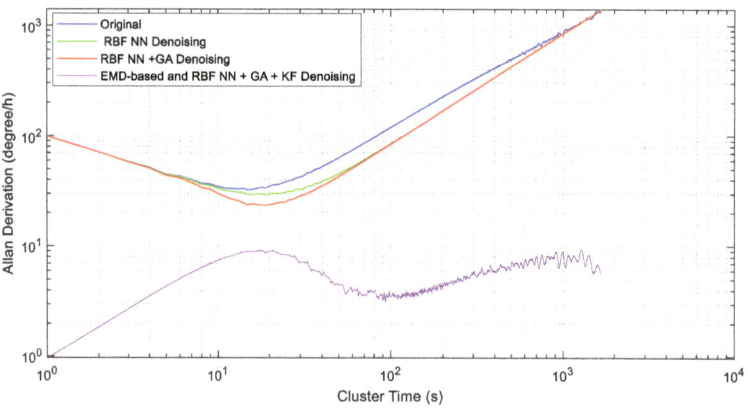

Figure 15. Allan variance analysis of the gyroscope's output after error processing.

Table 2. Allan variance analysis.

	Denoising						Temperature Compensation	
Original Data		RBF NN			RBF NN+GA		RBF NN+GA+KF	
$B(°/h)$	$N(°/h/Hz^{1/2})$	$B(°/h)$	$N(°/h/Hz^{1/2})$	$B(°/h)$	$N(°/h/Hz^{1/2})$		$B(°/h)$	$N(°/h/Hz^{1/2})$
34.66	99.608	28.42	99.6037	23.65	99.6011		3.589	0.967814

6. Conclusions

This paper studies the design and analysis of a new four-mass vibration MEMS gyroscope (FMVMG). The temperature compensation algorithm of the MEMS gyroscope is also studied in detail. First, the structure is simulated by finite element analysis software. The simulation results show that the FMVMG has two working modes: a driving mode and a sensing mode. The resonant frequency of the driving mode is 9.6096 kHz, the resonant frequency of the sensing mode is 9.6154 kHz, and the frequency interval between those two modes is 5.8 Hz. In the temperature compensation, a new fusion algorithm based on empirical mode decomposition (EMD) and combining a radial basis function neural network (RBF NN), genetic algorithm (GA), and Kalman filter (KF), is proposed. First, the gyroscope output is decomposed into three components by EMD and the sample entropy (SE), which are the noise component, the mixed component, and the drift component. Second, to process these three components, the specific steps are as follows: the high-frequency noise component is directly removed, and the RBF NN+GA+KF algorithms are used to denoise the mixed components. Third, the trend component of the temperature sensor output is used to compensate for the drift component of the MEMS gyroscope. Finally, the signal is reconstructed to obtain the final compensated signal. The results show that this method not only removes the noise signal of the gyroscope effectively but also protects the effective signal of the gyroscope. Finally, the Allan variance coefficient shows the comparison of the method with the original data. The data show that if the gyroscope output is compensated by the EMD and RBF NN + GA with the KF fusion method, the bias stability is increased from $34.66°/h$ to $3.589°/h$. The angular random walk is reduced from $99.608°/h/Hz^{1/2}$ to $0.967814°/h/Hz^{1/2}$.

Author Contributions: Z.L. and Y.C. conceived and designed the experiments; K.C., H.C., Y.G. and G.W. performed the experiments; J.Y. and Z.L. analysed the data; Z.L. wrote the paper. All authors have read and agreed to the published version of the manuscript.

Funding: This work is supported by the National Key Research and Development Program of China (No. 2022YFB3205000), the National Natural Science Foundation of China, the NSAF (Grant No. U2230206), the Technology Field Fund of Basic Strengthening Plan of China (2020JCJQJJ409, 2021-JCJQ-JJ-0315), the National Defense Basic Scientific Research Program (WDZC20190303), and the Pre-Research Field Foundation of Equipment Development Department of China (No. 61405170104 and No. 80917010501). The research is also supported by the Fundamental Research Program of Shanxi Province (20210302123020 and 20210302123062), the Shanxi Province Key Laboratory of Quantum Sensing and Precision Measurement (201905D121001), the Key Research and Development (R&D) Projects of Shanxi Province (202003D111004), the Beijing Key Laboratory of High Dynamic Navigation Technology Open Founding (HDN2021102), and the Fund for Shanxi "1331 Project" Key Subjects Construction.

Data Availability Statement: Not applicable.

Conflicts of Interest: The authors declare no conflict of interest.

References

1. Pribanić, T.; Petković, T.; Đonlić, M. 3D registration based on the direction sensor measurements. *Pattern Recognit.* **2019**, *88*, 532–546. [CrossRef]
2. Davidson, P.; Virekunnas, H.; Sharma, D.; Piché, R.; Cronin, N. Continuous Analysis of Running Mechanics by Means of an Integrated INS/GPSDevice. *Sensors* **2019**, *19*, 1480. [CrossRef]
3. Guo, Y.; Wang, P.; Ma, G.; Wang, L. Envelope oriented singularity robust steering law of control moment gyros for spacecraft attitude maneuver. *Trans. Inst. Meas. Control* **2019**, *41*, 954. [CrossRef]
4. Li, Y.; Hu, H.; Zhou, G. Using Data Augmentation in Continuous Authentication on Smartphones. *IEEE Internet Things J.* **2018**, *6*, 628–640. [CrossRef]
5. Jauhiainen, M.; Puustinen, J.; Mehrang, S.; Ruokolainen, J.; Holm, A.; Vehkaoja, A.; Nieminen, H. Identification of Motor Symptoms Related to Parkinson Disease Using Motion-Tracking Sensors at Home (KAVELI): Protocol for an Observational Case-Control Study. *JMIR Res. Protoc.* **2019**, *8*, e12808. [CrossRef] [PubMed]
6. Voicu, R.A.; Dobre, C.; Bajenaru, L.; Ciobanu, R.I. Human Physical Activity Recognition Using Smartphone Sensors. *Sensors* **2019**, *19*, 458. [CrossRef] [PubMed]
7. Yu, X.; Ma, H.; Jin, Z. Improving thermal stability of a resonator fiber optic gyro employing a polarizing resonator. *Opt. Express* **2013**, *21*, 358. [CrossRef] [PubMed]

8. Gao, Z.; Zhang, Y.; Gao, W. Method to determine the optimal layer number for the quadrupolar fiber coil. *Opt. Eng.* **2014**, *53*, 084106. [CrossRef]
9. Wang, J.; Yu, Y.; Chen, Y.; Luo, H.; Meng, Z. Research of a double fiber Bragg gratings vibration sensor with temperature and cross axis insensitive. *Opt.—Int. J. Light Electron Opt.* **2015**, *126*, 749–753. [CrossRef]
10. Yan, Y.; Ma, H.; Jin, Z. Reducing polarization-fluctuation induced drift in resonant fiber optic gyro by using single-polarization fiber. *Opt. Express* **2015**, *23*, 2002–2009. [CrossRef]
11. Ling, W.; Li, X.; Xu, Z.; Wei, Y. A Dicyclic Method for Suppressing the Thermal-induced Bias Drift of I-FOGs. *IEEE Photonics Technol. Lett.* **2015**, *28*, 272–275. [CrossRef]
12. Qian, G.; Fu, X.C.; Zhang, L.J.; Tang, J.; Liu, Y.R.; Zhang, X.Y.; Zhang, T. Hybrid fiber resonator employing LRSPP waveguide coupler for gyroscope. *Sci. Rep.* **2017**, *7*, 41146. [CrossRef] [PubMed]
13. Zhang, Z.; Yu, F. Analysis for the thermal performance of a modified quadrupolar fiber coil. *Opt. Eng.* **2018**, *57*, 017109. [CrossRef]
14. Zhang, Y.; Wu, Y.; Xu, X.; Xi, X.; Wu, X. Research on the method to improve the vibration stability of vibratory cylinder gyroscopes under temperature variation. *Int. J. Precis. Eng. Manuf.* **2017**, *18*, 1813–1819. [CrossRef]
15. Fontanella, R.; Accardo, D.; Moriello, R.S.L.; Angrisani, L.; Simone, D.D. An Innovative Strategy for Accurate Thermal Compensation of Gyro Bias in Inertial Units by Exploiting a Novel Augmented Kalman Filter. *Sensors* **2018**, *18*, 1457. [CrossRef] [PubMed]
16. Cao, H.; Zhang, Y.; Shen, C.; Liu, Y.; Wang, X. Temperature Energy Influence Compensation for MEMS Vibration Gyroscope Based on RBF NN-GA-KF Method. *Shock. Vib.* **2018**, *2018*, 2830686. [CrossRef]
17. Vahrameev, E.I.; Galyagin, K.S.; Oshivalov, M.A.; Savin, M.A. Method of numerical prediction and correction of thermal drift of the fiber-opticgyro. *Izv. Vuzov. Priborostr.* **2017**, *60*, 32–38. [CrossRef]
18. Narasimhappa, M.; Nayak, J.; Terra, M.H.; Sabat, S.L. ARMA model based adaptive unscented fading Kalman filter for reducing drift of fiber optic gyroscope. *Sens. Actuators A Phys.* **2016**, *251*, 42–51. [CrossRef]
19. Zhang, Q.; Wang, L.; Gao, P.; Liu, Z. An Innovative Wavelet Threshold Denoising Method for Environmental Drift of Fiber Optic Gyro. *Math. Probl. Eng.* **2016**, *2016*, 9017481. [CrossRef]
20. Song, R.; Chen, X.; Shen, C.; Zhang, H. Modeling FOG Drift Using Back-Propagation Neural Network Optimized by Artificial Fish Swarm Algorithm. *J. Sens.* **2014**, *2014*, 276043. [CrossRef]
21. Han, X.; Hu, S.M.; Luo, H. A Simplified Model of the Compensation Method for the Thermal Bias of a Ring Laser Gyro. *Lasers Eng.* **2014**, *27*, 119–126.
22. Antonova, M.V.; Matveev, V.A. Model of error of A fiber-opyic gyro exposed to thermal and magnetic fields. *Her. Bauman Mosc. State Tech. Univ. Ser. Instrum. Eng.* **2014**, *3*, 73–80.
23. Zha, F.; Xu, J.; Li, J.; He, H. IUKF neural network modeling for FOG temperature drift. *J. Syst. Eng. Electron.* **2013**, *24*, 838–844. [CrossRef]
24. Prikhodko, I.; Trusov, A.; Shkel, A.M. Compensation of drifts in high-Q MEMS gyroscopes using temperature self-sensing. *Sens. Actuators A: Phys.* **2013**, *201*, 517–524. [CrossRef]
25. Chen, X.; Shen, C. Study on error calibration of fiber optic gyroscope under intense ambient temperature variation. *Appl. Opt.* **2012**, *51*, 3755. [CrossRef] [PubMed]
26. Wang, X.; Wu, W.; Fang, Z.; Luo, B.; Li, Y.; Jiang, Q. Temperature Drift Compensation for Hemispherical Resonator Gyro Based on Natural Frequency. *Sensors* **2012**, *12*, 6434–6446. [CrossRef] [PubMed]
27. Cai, Q.; Zhao, F.; Kang, Q.; Luo, Z.; Hu, D.; Liu, J.; Cao, H. A Novel Parallel Processing Model for Noise Reduction and Temperature Compensation of MEMS Gyroscope. *Micromachines* **2021**, *12*, 1285. [CrossRef] [PubMed]
28. Zhou, Y.; Cao, H.; Guo, T. A Hybrid Algorithm for Noise Suppression of MEMS Accelerometer Based on the Improved VMD and TFPF. *Micromachines* **2022**, *13*, 891. [CrossRef]
29. Zhou, B.; Zhang, T.; Yin, P.; Chen, Z.; Song, M.; Zhang, R. Innovation of flat gyro: Center Support Quadruple Mass Gyroscope. In Proceedings of the 2016 IEEE International Symposium on Inertial Sensors and Systems, Laguna Beach, CA, USA, 22–25 February 2016.
30. Trusov, A.A.; Schofield, A.R.; Shkel, A.M. Gyroscope architecture with structurally forced anti-phase drive-mode and linearly coupled anti-phase sense-mode. In Proceedings of the TRANSDUCERS 2009-2009 International Solid-State Sensors, Actuators and Microsystems Conference, Denver, CO, USA, 21–25 June 2009.

*micromachines*

Article

Virtual Coriolis-Force-Based Mode-Matching Micromachine-Optimized Tuning Fork Gyroscope without a Quadrature-Nulling Loop

Yixuan Wu [1,2], Weizheng Yuan [1,2], Yanjun Xue [1,2], Honglong Chang [1,2] and Qiang Shen [1,2,*]

1 School of Mechanical Engineering, Northwestern Polytechnical University, Xi'an 710072, China; yuanwz@nwpu.edu.cn (W.Y.)
2 MOE Key Laboratory of Micro and Nano Systems for Aerospace, Northwestern Polytechnical University, Xi'an 710072, China
* Correspondence: shenq@nwpu.edu.cn

Abstract: A VCF-based mode-matching micromachine-optimized tuning fork gyroscope is proposed to not only maximize the scale factor of the device, but also avoid use of an additional quadrature-nulling loop to prevent structure complexity, pick-up electrode occupation, and coupling with a mode-matching loop. In detail, a mode-matching, closed-loop system without a quadrature-nulling loop is established, and the corresponding convergence and matching error are quantitatively analyzed. The optimal straight beam of the gyro structure is then modeled to significantly reduce the quadrature coupling. The test results show that the frequency split is narrowed from 20 Hz to 0.014 Hz. The scale factor is improved 20.6 times and the bias instability (BI) is suppressed 3.28 times. The observed matching accuracy demonstrates that a mode matching system without a quadrature suppression loop is feasible and that the proposed device represents a competitive design for a mode-matching gyroscope.

Keywords: MEMS gyroscope; mode-matching; matching error; virtual Coriolis force

Citation: Wu, Y.; Yuan, W.; Xue, Y.; Chang, H.; Shen, Q. Virtual Coriolis-Force-Based Mode-Matching Micromachine-Optimized Tuning Fork Gyroscope without a Quadrature-Nulling Loop. *Micromachines* **2023**, *14*, 1704. https://doi.org/10.3390/mi14091704

Academic Editor: Ha Duong Ngo

Received: 19 July 2023
Revised: 11 August 2023
Accepted: 13 August 2023
Published: 31 August 2023

1. Introduction

A gyroscope is a device for detecting the rotation rate of its carrier, and is one of the most important sensors for applications in fields such as navigation and industrial control [1]. MEMS-based gyroscopes are small and can be batch-fabricated at a low cost [2] compared with other gyroscope technologies such as photonic gyroscopes [3], nuclear magnetic gyroscopes [4], or fiber-optic gyroscopes [5]. However, the measurement accuracy of MEMS gyroscopes is limited due to machining errors. To improve the precision of MEMS gyroscopes, numerous approaches have been used, including the amelioration of the fabrication process [6], control of the vibration mode [7], temperature compensation [8], and back-end signal processing [9]. To benefit from the mode matching technology, the performance of MEMS gyroscopes has significantly improved to achieve navigation-grade accuracy [6]. This technology mainly relies on the fact that Coriolis-induced displacement in the sense mode reaches a maximum when the frequency split between the drive and sense modes of the gyroscope is zero, which means that the scale factor of the gyroscope is optimal.

The common mode matching scheme can be divided into three approaches. The first is the structure topology design of the gyroscope to achieve inherent mode matching. A ring gyroscope with L-shaped spokes based on a non-uniform radial width was designed to suppress the frequency split from 180 Hz to 50 Hz [10,11]. An elliptic-shaped structure connecting a thick ring alternately with multiple thin rings achieved the mode matching of 43 ppm at a center frequency of 69 kHz [12,13]. The frequency split of a micro-machined gyroscope was reduced to sub-Hz values with a temperature stability of 0.05 Hz over a

130 °C temperature range by placing critical mechanical elements, including coupling springs, anchors, and suspensions, at the center of the resonator structure [14].

The second approach to reduce the frequency split is employing a fabrication trimming scheme to modify the mass distribution via selective polysilicon deposition, laser trimming, or a focus-ion process. For example, a focus-ion beam process was used to trim the geometry to achieve the mode matching of 0.02 Hz at a resonant frequency of about 6 kHz [15]. A quantitative mass deposition process guided by a modification equation for the ring structure was devised to reach a mode matching level of 0.08 Hz [16]. Femtosecond laser trimming was applied with a mass-stiffness decoupling design to narrow the frequency split to 0.4 Hz [17].

The last widely used approach is an electrostatic tuning scheme based on the electrostatic spring softening technology to achieve mode matching by tuning the resonant frequency of the sense mode with a feedback DC voltage, which is a function of the characteristic signal reflecting the frequency split. For this category of method, the key issue is how to extract the related signal to the frequency split. Currently, existing methods for extracting the frequency split can be divided into three types, including (1) offline calibration, (2) signal symmetry comparison, and (3) small-signal-induced phase relation analysis.

For the offline calibration method, according to the previously defined quantitative relations between the frequency split, tuning DC voltage and angular rate output based on an offline experimental test, a lookup table (LUT) was established to identify the appropriate tuning voltage during the actual running to achieve mode matching. The LUT was built to regulate the frequency split of 0.01 Hz and to achieve an overall output bias drift of 0.76°/s over a 90 °C range [18]. A three-layer back propagation neural network controller was constructed to achieve a frequency split of 0.3 Hz over the temperature range from −40 °C to 80 °C by fitting the tuning voltage beforehand with the temperature [19]. Mode matching based on this method can be achieved quickly. Nevertheless, it is not easy to deal with the in-run, long-term mode matching uncertainty induced by mechanical fatigue deformation, residual stress change, or unexpected environment disturbances, such as vibration and shock.

With respect to the signal symmetry comparison method, double-side signal characteristics, including amplitude symmetry or noise power spectrum symmetry, are usually extracted as the matching error reference to judge the degree of mode matching. The symmetry of the both-sides band power spectrum of thermal noise in the sense mode was calculated to establish the functional relation between the degree of mode matching and the symmetrical error, achieving the frequency splits of 1.5 Hz [20,21] and 0.28 Hz [22], respectively. For high-quality gyroscopes, the noise characteristics dominated by the interface circuit are susceptible to environment disturbances, such as space electromagnetic coupling and stray capacitances fluctuation, which leads to mode matching accuracy deterioration. Two amplitude values of the signals at the lower and upper sides of the resonant frequency were compared by design-tuning the control loop to detect an amplitude difference, which will theoretically converge to zero when the two modes are matched. Nevertheless, due to the intrinsic, double-sided and non-uniform characteristic of high-Q gyroscopes, the slight asymmetry response was not fully eliminated [23].

With respect to the phase relation methods, the quadrature error is a typical characteristic excitation signal for mode matching [24–26]. In this way, the phase difference of the quadrature signal in the pick-up signal of the sense mode, with respect to the displacement of the drive mode, is extracted to generate a control error signal for the tuning voltage, where a 90° phase difference constituting a 0 V control error indicates the realization of mode matching. Nevertheless, the response to the angular rate input can couple into the control error and lead to a matching error when the sensor is of working status.

Another improved phase relation method is developed based on the virtual Coriolis force (VCF) [27,28]. In this way, an external signal for simulating the Coriolis force with the corresponding frequency and phase information was loaded to special electrodes in the sense mode to excite the vibration of the sense mode. An induced pick-up signal was demodulated by the reference signal in the phase with Coriolis force to generate the control error signal, which trended to zero when the mode matching was reached. Since the VCF method is based on the closed-loop classical control theory, it is a real-time mode matching method, which avoids the calibration for each sensor to significantly reduce experimental complexity. More importantly, the phase relation between the VCF and the pick-up signal is only decided by the matching degree of the internal gyroscope, and it is not affected by external disturbances. In reported works, the VCF method is used to narrow the frequency split to 0.03 Hz and sub-0.1 Hz on the tuning fork gyroscope (TFG) [29] and disk resonator gyroscope (DRG) [30], respectively.

Nevertheless, according to the precondition of the effectiveness of this VCF method, the quadrature error of the sensor can be suppressed as much as possible, which is usually realized through an additional closed quadrature-nulling loop [29–31];otherwise, the residual quadrature error, referring to the frequency split information, is an output to deteriorate the phase relation and reduce the sensitivity improvement of the gyroscope [28]. However, the additional quadrature suppression loop results in the complex gyroscope control circuit and reduction in the scale factor due to the occupation of sense pick-up electrodes. More importantly, the feedback of a quadrature-nulling loop may be coupled with the feedback of the mode-matching loop, leading to the leakage of stiffness trimming from the axis of quadrature stiffness to the axis of the sense mode. For example, with respect to the tuning fork gyroscope (TFG), the coexistence of slide-film excitation and squeeze-film excitation is why the electrostatic force is generated two-dimensionally and further simultaneously trims the stiffness of quadrature coupling and the sense mode [23]. With respect to the disk resonator gyroscope (DRG), the coupling is caused by the interaction between the sense mode stiffness and the quadrature coupling stiffness, due to the continuous distribution of the stiffness of the rings [32].

Considering the drawbacks of the phase relation method based on an additional quadrature suppression loop for the circuit complexity, occupation of electrodes, and coupling between quadrature nulling and mode matching, a simplified VCF-based mode-matching gyroscope without a quadrature-nulling loop and with an optimized TFG structure to achieve a sufficient structural quadrature suppression for mode matching is presented in this work. The proposed gyroscope without a quadrature-nulling loop has advantages in two aspects. On the one hand, as the quadrature-nulling loop is removed, the electrodes used for quadrature calibration can be used for detecting the sense mode displacement, and thus, the scale factor is maximized. On the other hand, the absence of the quadrature feedback means that the stiffness axis of the sense mode is only acted on by the frequency tuning voltage, and thus, the mode matching can be maintained more stably.

This work is organized as follows. The corresponding matching error caused by the quadrature error is theoretically analyzed in Section 2. Then, the optimization of the TFG structure is proposed and the effect of quadrature suppression is simulated. Third, the designed mode matching system with the estimated parameters of the gyroscope is simulated.

2. Theoretical Analysis of VCF-Based Mode Matching and Matching Error

2.1. Mode Matching System Design

Figure 1 shows the whole VCF-based mode matching schematic consisting of the mechanical part and electrical part. In the mechanical part, TFG is the object of the presented scheme, as shown in Figure 1b. In detail, TFG mainly consists of a drive, couple, and sense frames, which are marked by blue, green, and red, respectively. Due to the widely known tuning fork topology, the description of the Coriolis effect principle of TFG is omitted here. The electrical part contains three modules. The first module is the vibration-stabilizing closed-loop circuit for the drive mode, which is marked in blue. The second one is the equivalent electrical signal of the induced virtual Coriolis force, which is marked in red. The last one is the presented mode-matching circuit in the sense mode, which is marked in orange. In the mode-matching circuit, the total force, F_{tot}, consisting of Coriolis force, F_c, virtual Coriolis force, F_{vc}, and quadrature force, F_q, drives the sense mode to move. This displacement of sense mode y is then used as a voltage signal, V_{sp}, by the interface circuit. Then, the V_{sp} is divided into circuit units for angular an velocity output and a mode-matching, closed loop. In the mode-matching loop, V_{sp} is demodulated by the in-phase demodulation signal, V_{dm}, to obtain V_{sdm} as a function of the phase shift, φ_s^d, between F_{tot} and y. Through a low-pass filter, a comparator, and a proportional-integral (PI) controller, DC tuning voltage, V_t, is obtained and fed into the frequency tuning electrode of the sense mode. Based on the electrostatic softening effect, the resonant frequency of the sense mode, ω_s, is tuned by V_t and continuously varied until it equals to the resonant frequency of the drive mode ω_d. The detailed deduction is given as follows.

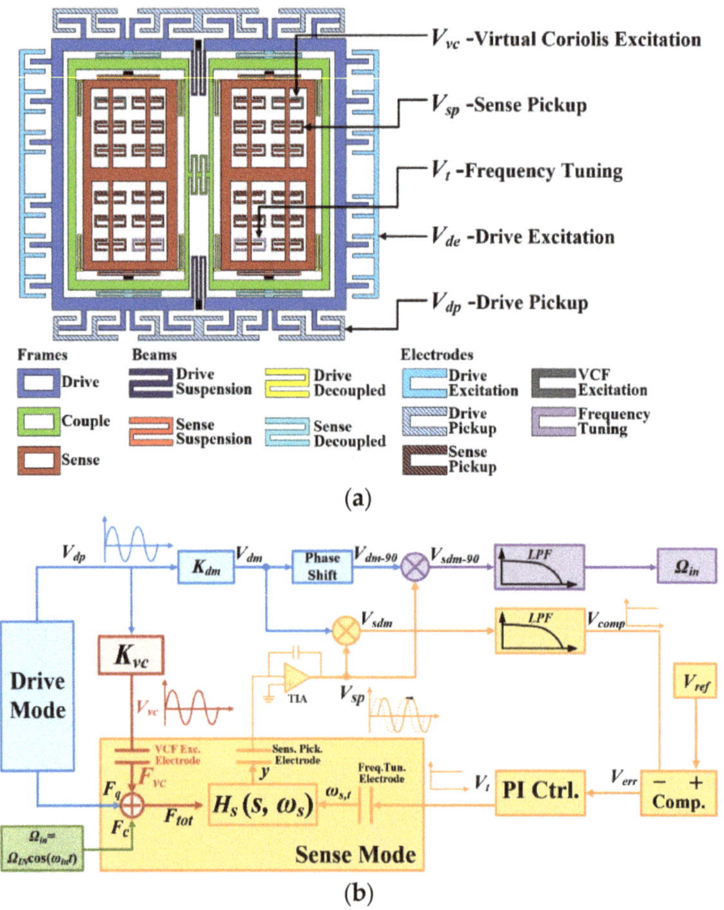

Figure 1. Schematic of (**a**) mode matching system with (**b**) the topology of the gyro sensor.

First, F_c, F_{vc}, and F_q, which are components of F_{tot}, are defined in Equation (1). F_c is the mechanical Coriolis force induced by the Coriolis effect under the angular rate input, Ω_{in}, with a frequency of ω_{in}. The amplitude of Coriolis force, F_c, is proportional to that of the angular rage input, Ω_{IN}. F_{vc} is the electrostatic force produced by the VCF excitation voltage, V_{vc}, which is obtained by amplifying the pick-up signal of the drive mode, V_{dp}, with the fixed gain, K_{vc}. According to the Coriolis effect theory and the principle of charge sensitive amplifier (CSA), all F_c, V_{dp}, V_{vc}, and F_{vc} have in-phase relations. Thus, F_{vc} is considered as artificial Coriolis force caused by the DC ($\omega_{in}= 0$) angular rate input, and it is defined as the VCF. The amplitudes of F_C, F_{VC}, and F_Q are considered as fix values due to the closed-loop amplitude control in the drive mode.

$$\begin{cases} F_c = F_C sin(\omega_d t)cos(\omega_{in} t) \\ F_{vc} = F_{VC} sin(\omega_d t) \\ F_q = F_Q cos(\omega_d t) \end{cases} \tag{1}$$

Then, the transfer function from F_{tot} to the pick-up signal V_{sp} is expressed as follows:

$$H_s(s) = \frac{\mathcal{L}\left[V_{sp}(t)\right]}{\mathcal{L}\left[F_{tot}(t)\right]} = \frac{K_{sp}/m_s}{s^2 + \frac{\omega_s}{Q_s}s + \omega_s^2} \tag{2}$$

where K_{sp} is the gain from y to V_{sp}, m_s is the mass of the sense mode, ω_s is the resonant frequency of the sense mode, and Q_s is the quality factor of the sense mode. According to the theory of the second-order system, V_{sp} in Equation (2) can be solved as follows:

$$\begin{cases} V_{sp} = \underbrace{\frac{F_C}{2}\left[G_s^{d+in}sin\left(\omega_{d+in}t + \varphi_s^{d+in}\right) + G_s^{d-in}sin\left(\omega_{d-in}t + \varphi_s^{d-in}\right)\right]}_{response\ to\ F_c} \\ \qquad\quad \underbrace{+F_{VC}G_s^d sin\left(\omega_d t + \varphi_s^d\right)}_{response\ to\ F_{vc}} + \underbrace{F_Q G_s^d sin\left(\omega_d t + \varphi_s^d + \frac{\pi}{2}\right)}_{response\ to\ F_q} \\ \omega_{d\pm in} \quad = \omega_d \pm \omega_{in} \end{cases} \tag{3}$$

$$\begin{cases} G_s^d = \dfrac{K_{sp}}{m_s\sqrt{\dfrac{\omega_d^2(\Delta\omega+\omega_d)^2}{Q_s^2}+\left[(\Delta\omega+\omega_d)^2-\omega_d^2\right]^2}} \\ \varphi_s^d = -arctan\left[\dfrac{\omega_d(\Delta\omega+\omega_d)/Q_s}{(\Delta\omega+\omega_d)^2-\omega_d^2}\right] \end{cases} \tag{4}$$

$$\begin{cases} G_s^{d+in} = \dfrac{K_{sp}}{m_s\sqrt{\dfrac{\omega_{d+in}^2(\Delta\omega+\omega_d)^2}{Q_s^2}+\left[(\Delta\omega+\omega_d)^2-\omega_{d+in}^2\right]^2}} \\ \varphi_s^{d+in} = -arctan\left[\dfrac{\omega_{d+in}(\Delta\omega+\omega_d)/Q_s}{(\Delta\omega+\omega_d)^2-\omega_{d+in}^2}\right] \end{cases} \tag{5}$$

$$\begin{cases} G_s^{d-in} = \dfrac{K_{sp}}{m_s\sqrt{\dfrac{\omega_{d-in}^2(\Delta\omega+\omega_d)^2}{Q_s^2}+\left[(\Delta\omega+\omega_d)^2-\omega_{d-in}^2\right]^2}} \\ \varphi_s^{d-in} = -arctan\left[\dfrac{\omega_{d-in}(\Delta\omega+\omega_d)/Q_s}{(\Delta\omega+\omega_d)^2-\omega_{d-in}^2}\right] \end{cases} \tag{6}$$

In Equation (3), V_{sp} comprises three terms, including F_c, F_{vc}, and F_q. In detail, the gain G_s and the phase shift φ_s from F_{tot} to V_{sp} can be deduced through the functions about frequency split $\Delta\omega$, where $\Delta\omega = \omega_s - \omega_d$. In the mode-matching loop, in order to extract the voltage amplitude, V_{sp} is demodulated by $V_{dm} = V_{DM}sin(\omega_d t)$, which is derived from V_{dp} multiplied by the fixed gain K_{dm}. V_{sdm} is deduced as follows:

$$\begin{aligned} V_{sdm}\left(\varphi_s^d\right) = {}& V_{sp}V_{dm} \\ = {}& \frac{V_{DM}}{2}\Bigg\{\frac{F_C G_s^{d+in}}{2}\left[cos\left(\omega_{in}t + \varphi_s^{d+in}\right) - cos\left(2\omega_d t + \omega_{in}t + \varphi_s^{d+in}\right)\right] \\ & +\frac{F_C G_s^{d-in}}{2}\left[cos\left(\omega_{in}t - \varphi_s^{d-in}\right) - cos\left(2\omega_d t - \omega_{in}t + \varphi_s^{d-in}\right)\right] \\ & +\frac{F_{VC}G_s^d}{2}\left[cos\left(\varphi_s^d\right) - cos\left(2\omega_d t + \varphi_s^d\right)\right] \\ & +\frac{F_Q G_s^d}{2}\left[-sin\left(\varphi_s^d\right) + sin\left(2\omega_d t + \varphi_s^d\right)\right]\Bigg\} \end{aligned} \tag{7}$$

From Equation (7), it can be seen that V_{sdm} contains the signals of three classes of components distinguished by frequency: the DC signals about φ_s^d, the low-frequency signals with frequency of ω_{in}, and the double-frequency signals with frequency of $2\omega_d$. Then, a low-pass filter (LPF) is designed after V_{sdm} reaching the DC component V_{comp}:

$$V_{comp}\left(\varphi_s^d\right) = \frac{V_{DM}G_s^d}{2}\left[(F_C + F_{VC})cos\left(\varphi_s^d\right) - F_Q sin\left(\varphi_s^d\right)\right] \tag{8}$$

Herein, V_{comp} is only related to the phase shift φ_s^d. Subsequently, a subtracter with negative input, V_{comp}, and a positive input, $V_{ref} = 0$, are designed to obtain the error signal V_{err}:

$$\begin{aligned}V_{err}\left(\varphi_s^d\right) &= V_{ref} - V_{comp}\\ &= -\frac{V_{DM}G_s^d}{2}\left[(F_C + F_{VC})cos\left(\varphi_s^d\right) - F_Q sin\left(\varphi_s^d\right)\right]\end{aligned} \tag{9}$$

V_{err} is then input into the PI controller to pursue the DC tuning voltage, V_t, which is fed into the frequency-tuning electrode of the sense mode. According to electrostatic spring softening theory [33,34], the stiffness variation of the sense mode k_t versus tuning voltage V_t can be expressed as follows:

$$\begin{cases} k_t(V_t) = \varepsilon_0 HL\left(\frac{1}{D_1^3} + \frac{1}{D_2^3}\right)\left[N_t(V_p - V_t)^2 + (N_{sp} + N_{vc})V_p^2\right] \\ \omega_s(V_t) = \sqrt{\frac{k_{s0} - k_t}{m_s}} \end{cases} \tag{10}$$

The design parameters of the sensor structure in Equation (10) are defined in Table 1.

Table 1. Parameters of the sensor structure.

Symbol	Description
ε_0	vacuum dielectric constant
m_s	mass of the sense mode
k_s	$k_s = \omega_s^2 m_s$, stiffness of sense mode
k_{s0}	stiffness of the sense mode when $V_t = 0$
H	thickness of the device layer
L	overlap length of the combs
D_1	upper gap of the electrodes
D_2	lower gap of the electrodes
N_{sp}	number of capacitor pairs for the sense pick-up
N_{vc}	number of capacitor pairs for the VCF feedback
N_t	number of capacitor pairs for frequency tuning

The process reflected by Equations (3)−(10) evolves continuously until V_{err} equals to zero, which is proved by Equations (11)−(14) in the following Section 2.2. In this case, if F_Q is suppressed and can be neglected, φ_s^d satisfies the relation $\varphi_s^d = \pi/2$ according to Equation (9). Furthermore, according to Equation (4), it is deduced that $\Delta\omega = 0$, suggesting that mode matching is realized. In practice, F_Q is not strictly controlled and it produces matching errors within the system. Thus, the influence of F_Q is analyzed in Section 2.2.

2.2. Quantification of Matching Errors

As mentioned above, the designed system is self-stabilizing and requires quadrature suppression for mode matching. However, the quadrature force is not strictly controlled with the absence of the quadrature-suppression circuit in this system for maximizing the detection sensitivity of V_{sp}. Herein, the matching error caused by F_Q is mathematically analyzed.

As the mode matching system is designed with stability conditions, it can be linearized around its stable point to study the matching error in a stable state. In this case, the equivalent system concerning how $\Delta\omega$ is established, as shown in Figure 2, and the nonlinear relations between two nets are linearized. For example, the square root relation between k_s and ω_s is replaced by the fixed gain, G_ω.

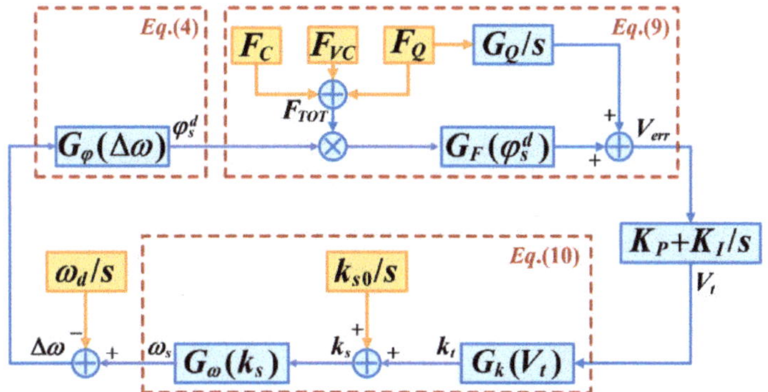

Figure 2. Schematic of the frequency split controlling.

By taking V_{err}, V_t, and $\Delta\omega$ as the output variables, three closed-loop transfer functions are constructed:

$$\begin{cases} \left\{\left[\mathcal{L}(V_{err})\left(K_P+\frac{K_I}{s}\right)G_k+\frac{k_{s0}}{s}\right]G_\omega-\frac{\omega_d}{s}\right\}G_\varphi F_{TOT}G_F+\frac{F_QG_Q}{s}=\mathcal{L}(V_{err}) \\ \left\{\left[\left(\mathcal{L}(V_t)G_k+\frac{k_{s0}}{s}\right)G_\omega-\frac{\omega_d}{s}\right]G_\varphi F_{TOT}G_F+\frac{F_QG_Q}{s}\right\}\left(K_P+\frac{K_I}{s}\right)=\mathcal{L}(V_t) \\ \left[\left(\mathcal{L}(\Delta\omega)G_\varphi F_{TOT}G_F+\frac{F_QG_Q}{s}\right)\left(K_P+\frac{K_I}{s}\right)G_k+\frac{k_{s0}}{s}\right]G_\omega-\frac{\omega_d}{s}=\mathcal{L}(\Delta\omega) \end{cases} \tag{11}$$

where $F_{TOT}=F_C+F_{VC}+F_Q$ and $\mathcal{L}(*)$ is the Laplace operator. By solving Equation (11), the Laplace transformations of V_{err}, V_t, and $\Delta\omega$ are obtained as follows:

$$\begin{cases} \mathcal{L}(V_{err})=-\dfrac{G_QF_Q+G_\varphi G_F F_{TOT}(G_\omega k_{s0}-\omega_d)}{(G_kG_\omega G_\varphi G_F K_P F_{TOT}-1)s+G_kG_\omega G_\varphi G_F K_I F_{TOT}} \\ \mathcal{L}(V_t)=-\dfrac{(K_I+K_Ps)\left[G_QF_Q+G_\varphi G_F F_{TOT}(G_\omega k_{s0}-\omega_d)\right]}{(G_kG_\omega G_\varphi G_F K_P F_{TOT}-1)s^2+G_kG_\omega G_\varphi G_F K_I F_{TOT}s} \\ \mathcal{L}(\Delta\omega)=-\dfrac{G_kG_\omega G_Q K_I F_Q+s\left(G_kG_\omega G_Q K_P F_Q+G_\omega k_{s0}-\omega_d\right)}{(G_kG_\omega G_\varphi G_F K_P F_{TOT}-1)s^2+G_kG_\omega G_\varphi G_F K_I F_{TOT}s} \end{cases} \tag{12}$$

By conducting an inverse Laplace transformation for Equation (12), the solutions in the time domain of V_{err}, V_t, and $\Delta\omega$ are solved as follows:

$$\begin{cases} V_{err}=-\dfrac{G_QF_Q+G_\varphi G_F F_{TOT}(G_\omega k_{s0}-\omega_d)}{G_kG_\omega G_\varphi G_F K_P F_{TOT}-1}e^{-\frac{1}{T}t} \\ V_t=-\dfrac{G_QF_Q+G_\varphi G_F F_{TOT}(G_\omega k_{s0}-\omega_d)}{G_kG_\omega G_\varphi G_F F_{TOT}}-\dfrac{G_QF_Q+G_\varphi G_F F_{TOT}(G_\omega k_{s0}-\omega_d)}{G_kG_\omega G_\varphi G_F F_{TOT}\left(G_kG_\omega G_\varphi G_F K_P F_{TOT}-1\right)}e^{-\frac{1}{T}t} \\ \Delta\omega=-\dfrac{G_QF_Q}{G_\varphi G_F F_{TOT}}+\left[\dfrac{G_QF_Q}{G_\varphi G_F F_{TOT}}-\dfrac{G_kG_\omega G_Q K_P F_Q+G_\omega k_{s0}-\omega_d}{G_kG_\omega G_\varphi G_F K_P F_{TOT}-1}\right]e^{-\frac{1}{T}t} \end{cases} \tag{13}$$

where $1/T=G_kG_\omega G_\varphi G_F K_I F_{TOT}/\left(G_kG_\omega G_\varphi G_F K_P F_{TOT}-1\right)$ is the damping coefficient. With increasing the time, t, the exponential terms in Equation (13) converge to zero and the stable points of V_{err}, V_t, and $\Delta\omega$ converge to the following equations:

$$\begin{cases} V_{err}=0 \\ V_t=-\dfrac{G_QF_Q}{G_kG_\omega G_\varphi G_F F_{TOT}}-\dfrac{G_\omega k_{s0}-\omega_d}{G_kG_\omega} \\ \Delta\omega=-\dfrac{G_QF_Q}{G_\varphi G_F F_{TOT}} \end{cases} \tag{14}$$

Obviously, the stable point of V_{err} is always equal to zero. This feature is provided by the principle of the PI controller and not affected by F_Q. In contrast, the stable point of $\Delta\omega$ is a residual term about F_Q, which causes the matching error of the system. Equation (14) indicates $V_{err} = 0$ when the system is stable, and the phase shift, φ_s^d, can be deduced using Equation (9):

$$\varphi_s^d = arctan\left(\frac{F_C + F_{VC}}{F_Q}\right) \tag{15}$$

By combining Equations (4) and (15), the stable solution of $\Delta\omega$ is expressed as follows:

$$\Delta\omega = \left\{\frac{F_Q}{2Q_s(F_C + F_{VC})} + \sqrt{\left[\frac{F_Q}{2Q_s(F_C + F_{VC})}\right]^2 + 1} - 1\right\}\omega_d \tag{16}$$

According to Equation (16), the matching error is affected by $F_C + F_{VC}$ and F_Q. Specifically, the matching error is proportional to the ratio of F_Q to $F_C + F_{VC}$ when it satisfies Equation (17):

$$\left[\frac{F_Q}{2Q_s(F_C + F_{VC})}\right]^2 \ll 1 \tag{17}$$

Thus, $|F_C + F_{VC}|$ should be reduced as much as possible with respect to $|F_Q|$ for the matching error reduction. On the contrary, the increase in $|F_Q|$ will enlarge the matching error. In addition, according to Figure 2, $F_C + F_{VC}$ constitutes the loop gain of the mode matching system and its polarity should be constant in working to keep the negative feedback characteristics of the mode matching system; otherwise, if the value of $F_C + F_{VC}$ exceeds the zero point, the system can possibly oscillate. In summary, the design of the mode matching system turns to an optimization problem:

$$\begin{cases} max & |F_C + F_{VC}| \\ s.t. & F_C + F_{VC} > 0, \quad F_{VC} > 0 \end{cases} \tag{18}$$

In this system, F_{VC} is designed positively, with a fixed value when the drive loop of gyro is stable. Then, the value of $F_C + F_{VC}$ should be discussed in two cases:

- Case $F_C \geq 0$

The case that $F_C \geq 0$ corresponds to the situation where the positive direction angular rate is the input. In this case, $F_C + F_{VC}$ always satisfies the constraint condition in Equation (18) and it is increased with increasing the angular rate. Meanwhile, the matching error is decreased by the input angular rate, according to Equation (16).

- Case $F_C < 0$

The case that $F_C < 0$ corresponds to the situation when the negative direction angular rate is the input. In this case, we have $F_C + F_{VC} = F_{VC} - |F_C|$. To guarantee the constraint condition in Equation (18), it is required that $|F_C| < F_{VC}$. Combined with Equation (16), it is revealed that the increment in the angular rate in the negative direction magnifies the matching error, and F_{VC} is the boundary value of the angular rate in negative direction.

The numerical analysis of the matching error with the designed range of F_C, F_{VC}, and F_Q is proposed. To explicitly compare F_C, F_{VC}, and F_Q, they are equivalently transformed into the form of amplitudes of angular rate Ω_{IN}, Ω_{VC}, and Ω_Q, respectively.

First, Ω_{IN} is the test condition provided by the rate table, and thus it is known as the setting value. In this paper, it varies from $-49°/s$ to $-50°/s$.

Then, the value of F_{VC} is selected because it is the constant excitation of the in-run mode matching system. F_{VC} is calculated according to the electrostatic effect of the mechanical capacitor.

$$F_{VC} = \varepsilon_0 N_{vc} HL\left(\frac{1}{D_1^2} - \frac{1}{D_2^2}\right)V_p V_{VC} \tag{19}$$

where V_{VC} is the amplitude of V_{vc}. Meanwhile, F_{VC} is also related to Ω_{VC} under the Coriolis effect.

$$F_{VC} = 2\Omega_{VC} m_s \dot{x} \tag{20}$$

where $\dot{x} = \omega_d x$ is the velocity of drive mode and x is the displacement of the drive mode. Combining Equations (19) and (20), the virtual angular rated input Ω_{VC} is expressed by Equation (21), as follows:

$$\Omega_{VC} = \frac{\varepsilon_0 N_{vc} H L}{2 m_s \dot{x}} \left(\frac{1}{D_1^2} - \frac{1}{D_2^2} \right) V_p V_{VC} \tag{21}$$

In this paper, Ω_{VC} is $50°/s$ via trimming V_{VC}.

Third, Ω_Q is evaluated by the scale factor of the gyro. Considering that Ω_{in} is set as a DC rotation, it is obvious that

$$\frac{\Omega_Q}{\Omega_{IN}} = \frac{V_Q}{V_{IN}} \tag{22}$$

where V_Q and V_{IN} are the DC voltages induced by F_Q and F_C at the output of the sense mode detection circuit, respectively. As V_{IN}/Ω_{IN} is the scale factor of the gyro, Ω_Q is finally equal to the ratio of V_Q to the scale factor. Experimentally, Ω_Q can be estimated under $15°/s$ for the proposed gyro. Furthermore, the numerical relations between Ω_{IN}, Ω_Q, and Δf is calculated through Equation (16), where F_C, F_{VC}, and F_Q are replaced by Ω_{IN}, Ω_{VC}, and Ω_Q, respectively. Additionally, $\Delta f = \Delta \omega / 2\pi$ represents the matching error in unit of Hertz. Calculation results are shown in Figure 3. Figure 3a shows the continuous numerical relationship between Δf and Ω_Q with several samples of F_C, and Figure 3b corresponds to the opposite case. It should be noted that Δf increases significantly when Ω_{IN} approaches the boundary negative direction $(-\Omega_{VC})$. Thus, the merge of a negative measurement is required for the specific matching error target in operation. By comparing the effects of Ω_Q and Ω_{IN} on Δf, the target of Δf is set as 0.5 Hz for the negative measurement limit $\Omega_{IN} = -35°/s$ when $\Omega_Q = 15°/s$, which is an empirical value in our previous design of TFG [35]. The negative measurement limit can be possibly expanded to $-45°/s$ when Ω_Q is further suppressed below $5°/s$, which is the estimated value referring to [36].

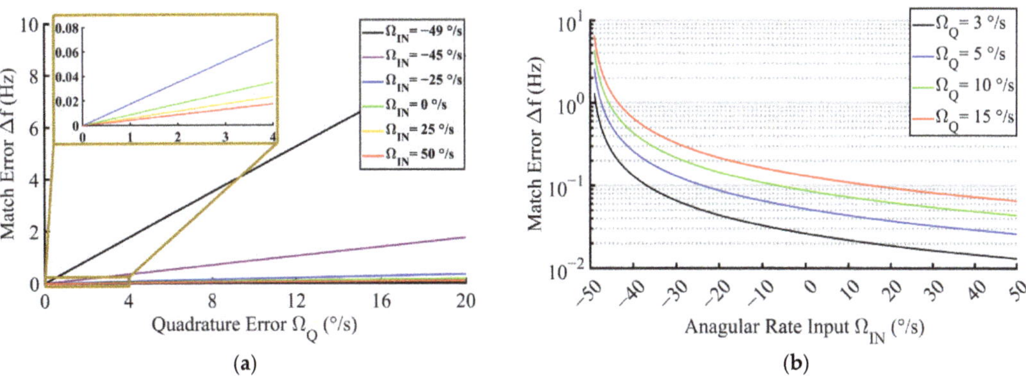

Figure 3. Matching error versus varying (**a**) quadrature error and (**b**) angular rate input.

3. System Implementation

3.1. Optimization of the Gyro Sensor

The optimized TFG structure compared with the original prototype is demonstrated in Figure 4. The folded sense suspension beams and the sense-decoupled beams are replaced by the straight ones, which provide better stiffness along the drive axis of the gyro, thereby improving the oscillation stability of a couple frame and the rigid constraints of a sense frame along the drive axis. Consequently, the quadrature couple between the drive and sense modes is further suppressed.

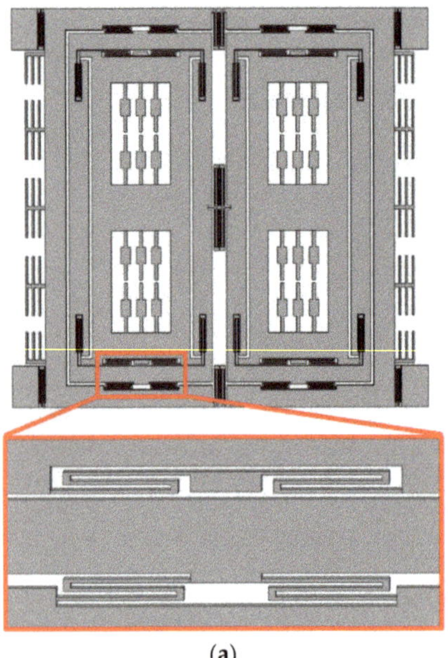

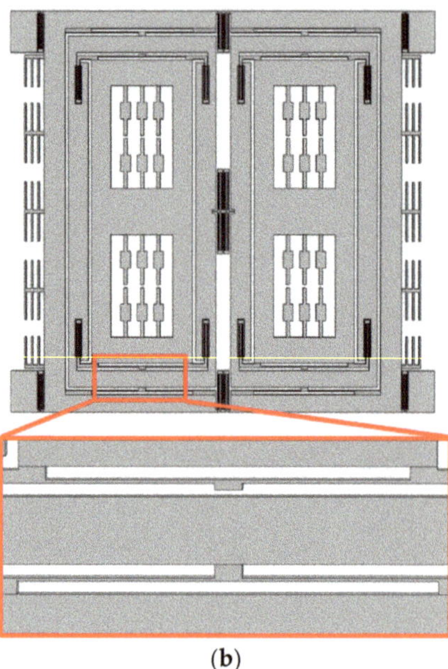

(a)　　　　　　　　　　　　　　　　　　**(b)**

Figure 4. Structure comparison of (**a**) original structure and (**b**) optimized structure.

The effect of quadrature suppression of TFG structures is verified through a mechanical simulation. The excitation for quadrature error is operated as shown in Figure 5. To produce the mode coupling, the deformation of the element mesh is used to drive the suspension beams for simulating the machining error, which induces the quadrature coupling between the drive and sense mode (Figure 5a). To produce the quadrature force, the drive mode is excited by adding a harmonic disturbance load to the drive excitation electrodes (Figure 5b). The frequency sweep of the harmonic disturbance is executed in the simulation and the displacement of the sense mode is recorded to evaluate the level of quadrature error.

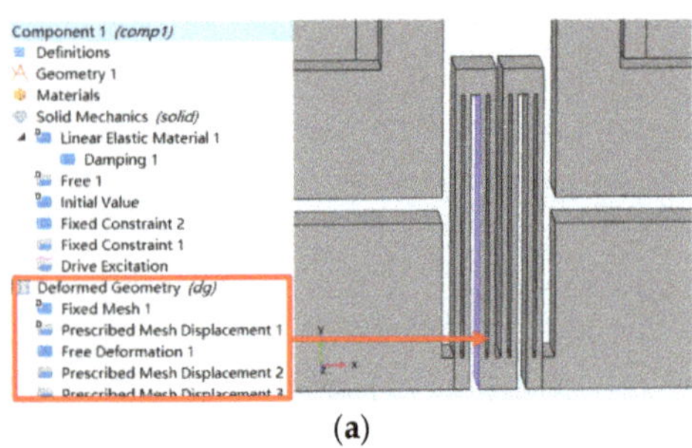

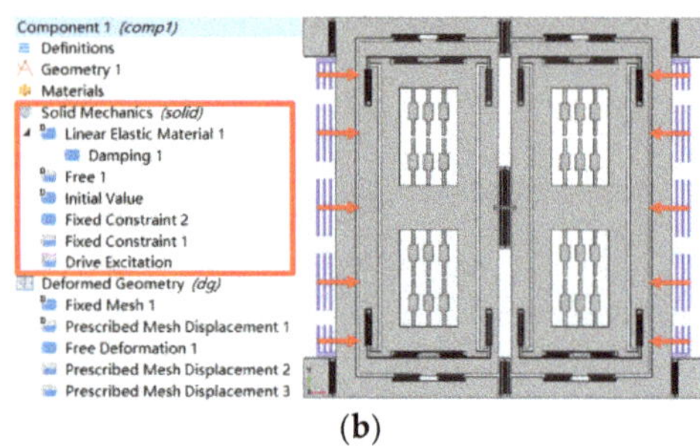

(a)　　　　　　　　　　　　　　　　　　**(b)**

Figure 5. Steps of quadrature error simulation (**a**) mesh deformation and (**b**) load definition.

The simulation results are shown in Figures 6 and 7. Figure 6 shows the distortion of the sense suspension beams caused by the quadrature coupling when the harmonic disturbance load is used for the drive excitation electrodes. By comparison, a significant distortion emerges on the sense suspension beams of the original structure, whereas the sense suspension beams of the optimized structure present a better resistance against the drive coupling. Figure 7 shows the frequency sweep of the displacement corresponding to the two structures of the quadrature coupling. The maximum amplitudes of displacement of the two structures are 0.53 um and 0.11 um, respectively. The quadrature errors Ω_Q of the two structures are equivalent to $17°/s$ and $3.5°/s$, respectively.

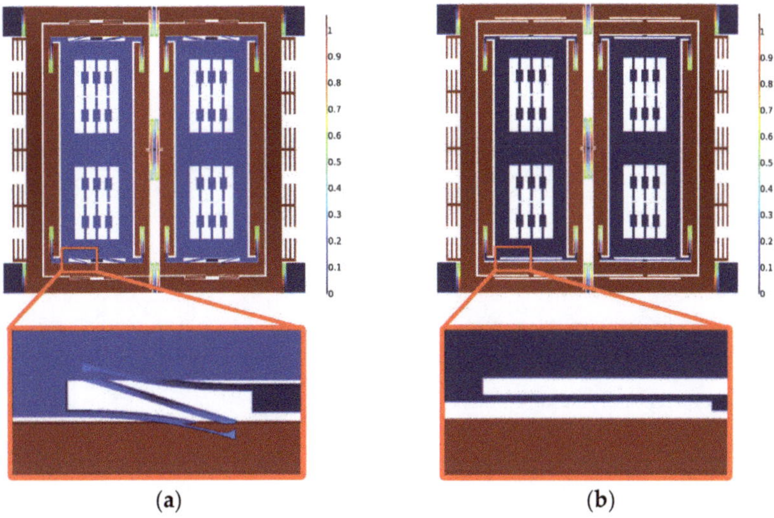

Figure 6. Quadrature couple simulation of (**a**) the original structure and (**b**) the optimized structure.

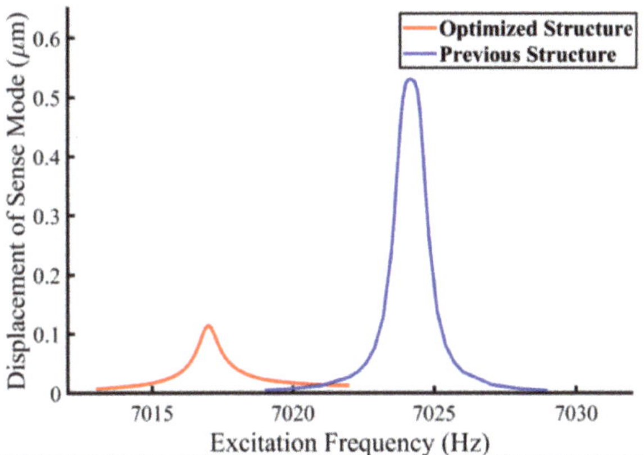

Figure 7. Comparison of coupled displacement in the sense mode.

3.2. Simulation System Construction

According to the designed system, a parameterized system model for the real circuit implementation is constructed as shown in Figure 8. The simulation parameters are summarized in Table 2. The simulation works consist of a transient response, stable matching error, and scale factor.

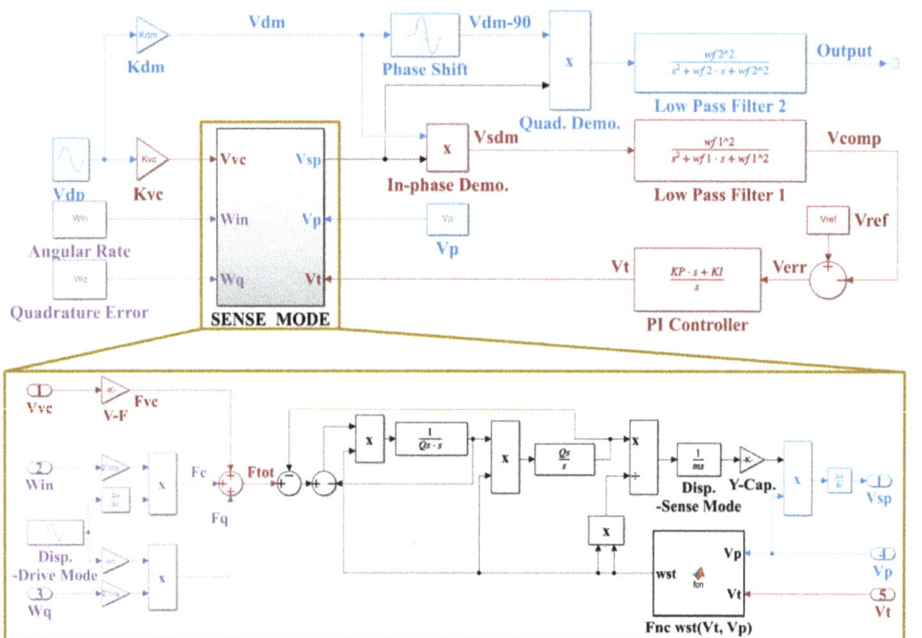

Figure 8. Schematic of the simulation model.

Table 2. Simulation parameters.

Parameter	Value
f_d	7000 Hz
f_s	6990 Hz
Q_s	8000
Ω_{IN}	$-45°/s \sim 50°/s$
Ω_{VC}	$50°/s$
Ω_Q	$5°/s \sim 15°/s$

The transient response of the mode matching system varies relatively to Ω_{IN}, as shown in Figure 9. The frequency split converges to zero during the full scale, from $-45°/s$ to $50°/s$, in the absence of the quadrature error. The setting time is about 0.6 s when $\Omega_{IN} = 50°/s$, whereas it is increased to about 11 s when Ω_{IN} gradually varies to $-45°/s$. This variation is caused by the closed-loop gain related to $\Omega_{IN} + \Omega_{VC}$ when the quadrature error is neglected according to the equivalent system in Figure 2.

The matching error caused by the quadrature error is simulated with a series of Ω_{IN} and Ω_Q. The results are summarized in Table 3. The relations between Δf, Ω_{IN}, and Ω_Q agrees well with the description in Section 2.2. The matching error is controlled to under 0.41 Hz within the measurement range of $\Omega_{IN} > -45°/s$ when Ω_Q is suppressed under $5°/s$.

The scale factor simulation is performed as shown in Figure 10. The discrete points are the simulated samples, and the solid lines are the corresponding first-order fitting lines. The fitted scale factors of the open-loop operation and closed-loop operation with the quadrature error varying from $0°/s$ to $15°/s$ are calculated as 0.983 mV/$°$/s, 16.095 mV/$°$/s, 16.082 mV/$°$/s, 16.089 mV/$°$/s, and 16.102 mV/$°$/s, respectively. The scale factor of the gyro with the mode matching system is improved 16.36 times on average, compared with that of the open-loop operation. Compared with the result in the case of $\Omega_Q = 0$, the worst deterioration of the scale factor caused by the quadrature error is 0.12%.

The nonlinearity of the open-loop operation and closed-loop operation with quadrature error varying from $0°/s$ to $15°/s$ is calculated as 0.056%, 0.042%, 0.079%, 0.084%, and 0.031%, respectively. The correlation between the nonlinearity and quadrature error cannot be observed through simulation. It can be considered that the proposed system without circuits for the quadrature suppression is feasible and acceptable with respect to both the scale factor improvement and the nonlinearity maintenance.

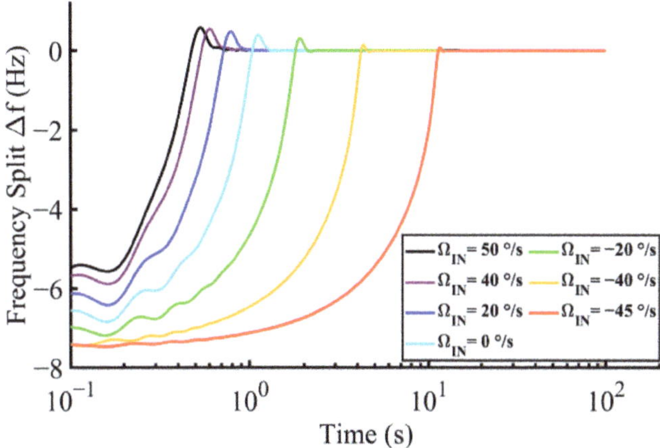

Figure 9. Transient response with different angular rate inputs.

Table 3. Matching error (Hz) versus quadrature error and angular rate input.

		\multicolumn{5}{c}{$\Omega_{IN}(°/s)$}				
		−45	−20	0	20	50
$\Omega_Q(°/s)$	5	0.41	0.09	0.06	0.05	0.04
	10	0.88	0.16	0.10	0.08	0.08
	15	1.36	0.15	0.15	0.11	0.09

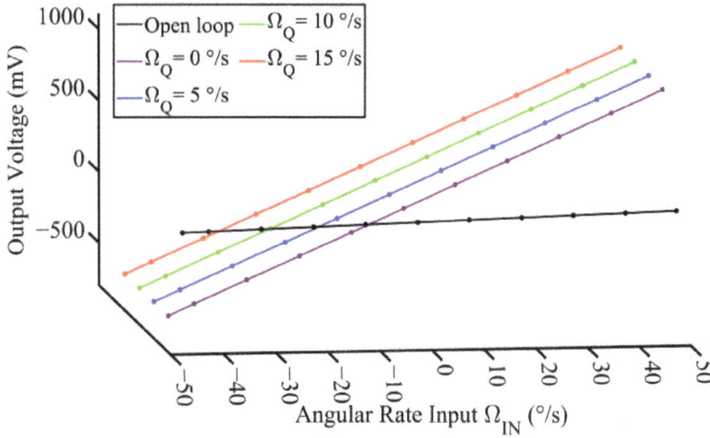

Figure 10. Scale factor simulation with different quadrature errors.

4. Experiment

4.1. TFG Sensor Fabrication

The fabrication of TFG sensor is divided into three parts: the through-silicon-via (TSV) process, the device process, and the wafer-level-package (WLP) process, as shown in Figure 11. The TSV process contains the following treatments: (a) etching the back side with TMAH; (b) deep reactive ion etching (DRIE) for trench; (c) filling trench with thermal oxide and in situ doped polysilicon; and (d) etching the cavity and removing the oxide layer. The device process contains the following treatments: (e) etching the shallow-bottom cavity and thermal oxidation; (f) vacuum bonding, grinding, and chemical mechanical polishing (CMP) the device layer; and (g) DRIE device wafer to release movable structures. The WLP process contains the following treatments: (h) vacuum-bonding the TSV wafer to the device wafer; (i) plasma-enhanced chemical vapor deposition (PECVD) of TEOS and reactive ion etching (RIE); (j) sputtering and dry-etching the metal layer; and (k) PECVD of TEOS and RIE.

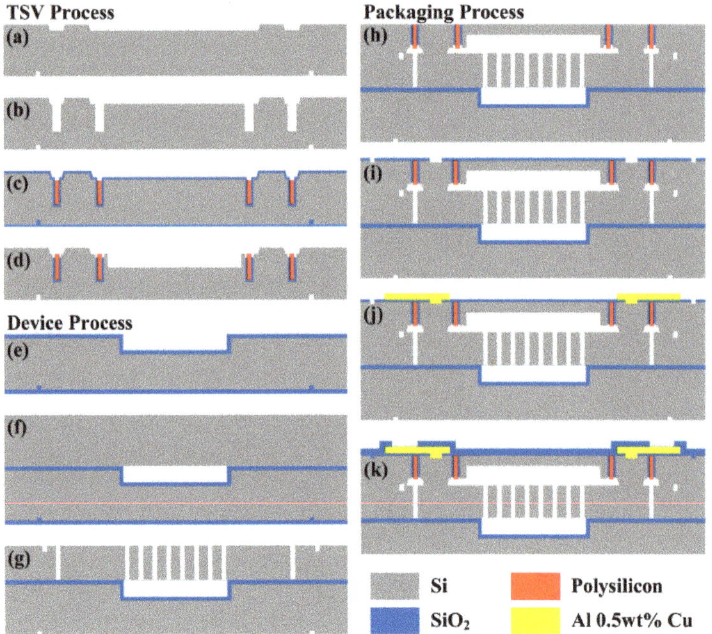

Figure 11. Fabrication process of TFG sensor: (**a**) etching the back side with TMAH; (**b**) deep reactive ion etching (DRIE) for trench; (**c**) filling trench with thermal oxide and in situ doped polysilicon; (**d**) etching the cavity and removing the oxide layer; (**e**) etching the shallow-bottom cavity and thermal oxidation, (**f**) vacuum bonding, grinding, and chemical mechanical polishing (CMP) the device layer; (**g**) DRIE device wafer to release movable structures; (**h**) vacuum-bonding the TSV wafer to the device wafer; (**i**) plasma-enhanced-chemical-vapor deposition (PECVD) of TEOS and reactive ion etching (RIE); (**j**) sputtering and dry-etching the metal layer; and (**k**) PECVD of TEOS and RIE.

4.2. TFG Sensor Tests

The implemented test board and TFG sensor are shown in Figure 12. The experiments include the open-loop sweep for the mode characteristics, the open-loop frequency tuning for verifying the electrostatic tuning effect, the transient response of drive mode for the demodulation signals, the start-up progress of the mode-matching circuit for evaluating the frequency split, and the scale factor, as well as the Allan variance for evaluating the overall performance.

Figure 12. PCB test board with the TFG sensor.

The sweep tests of the drive and sense modes were conducted under the open-loop operation. The magnitude–frequency response and the phase–frequency response of the two modes are shown in Figure 13. It can be found that the resonant frequencies of the drive and sense modes are 6960.09 Hz and 6940.28 Hz, respectively. Additionally, the quality factors of the two modes are calculated with the 3dB bandwidth, which are Q_d = 24,857 and Q_s = 7977, respectively.

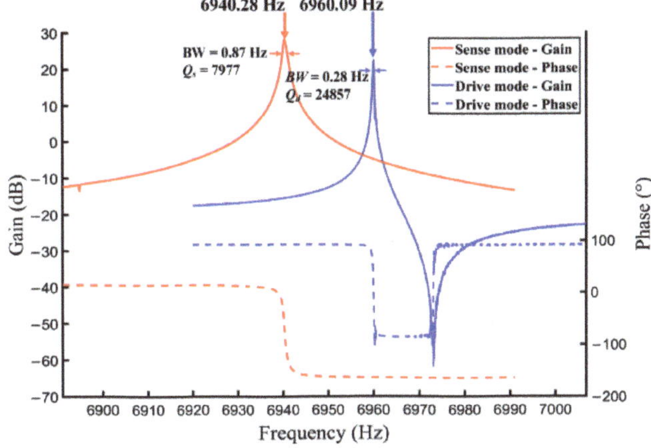

Figure 13. Sweep tests of the drive and sense modes under open-loop operation.

The open-loop frequency tuning test was conducted as shown in Figure 14. The resonant frequency of the drive mode is independent of the tuning voltage because of the fully decoupled structure. The resonant frequency of sense mode is firstly increased and then decreased when V_t exceeds 5 V. This is because an initial stiffness softening is induced by the bias voltage, V_p, and is then offset by the improvement in V_t according to Equation (10).

The transient response of the drive mode under the mode mismatching operation is shown in Figure 15. The drive mode proceeds about 3 s to reach a stable status. The drive excitation signal, V_{de}, is maintained at about 128 mV and the drive pickup signal V_{dp} is controlled at 1 V. The drive mode oscillates at its resonant frequency of 6960.3 Hz, which is indicated by the antiphase between V_{de} and V_{dp}. Furthermore, V_{dp} is reused as the in-phase demodulation signal V_{dm} and shifted with 90° to get the orthogonal demodulation signal V_{dm-90}, as shown in Figure 16.

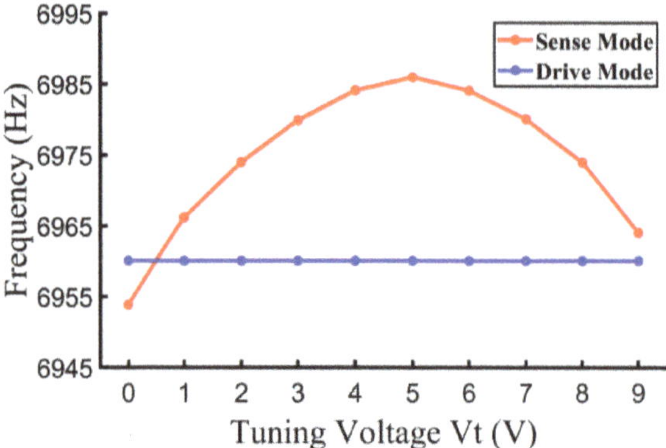

Figure 14. Open-loop frequency tuning test.

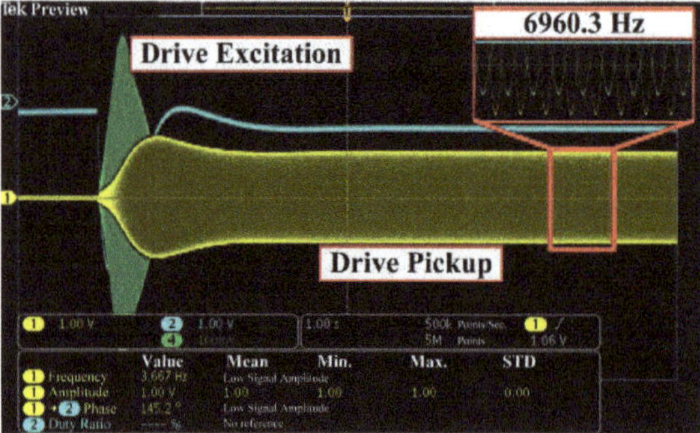

Figure 15. Transient response of the drive mode.

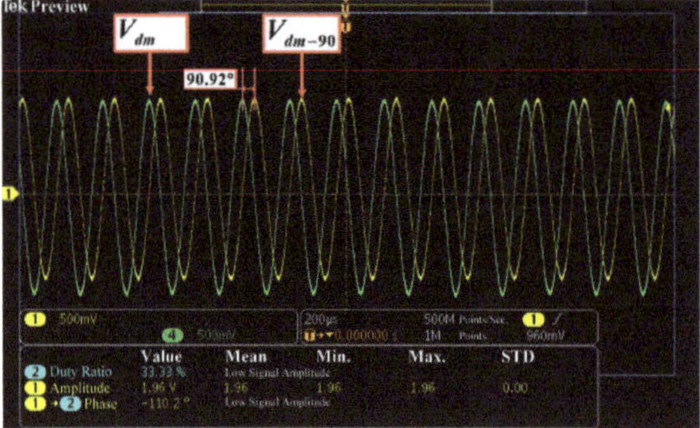

Figure 16. Phase shift of the demodulation signals.

The start-up progress of the mode-matching circuit and the oscillation of the sense mode are shown in Figure 17. In Figure 17a, the error voltage, V_{err}, and the tuning voltage, V_t, converge to 0 V and 715 mV after 2 s, respectively. The detected angular rate output voltage is 350 mV, which reflects the virtual angular rated input, Ω_{VC}. The zoomed waveform in Figure 17b shows the phase difference of 88.18° between V_{dp} and V_{sp}, which is close to the theory value of 90° when the gyro is under the mode matching operation. According to the relationship between the phase shift, φ_s^d, and frequency split, Δf, as expressed by Equation (4), Δf is controlled at about 0.014 Hz. It has been proven that without the additional circuits for quadrature suppression, the structure-optimized gyro sensor still provides an excellent performance of mode decoupling, and the residual quadrature error barely produces any effect on the mode matching.

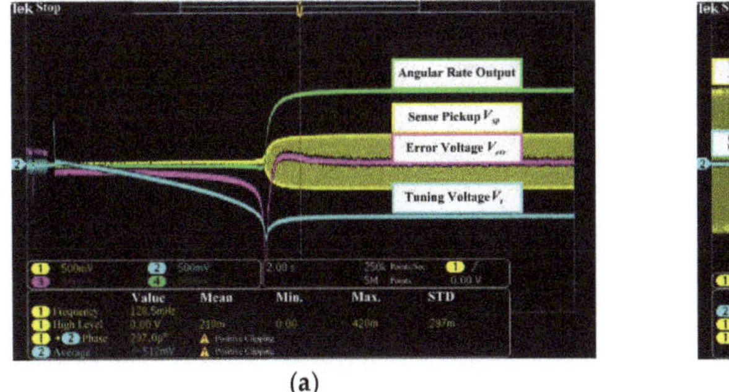

(a)

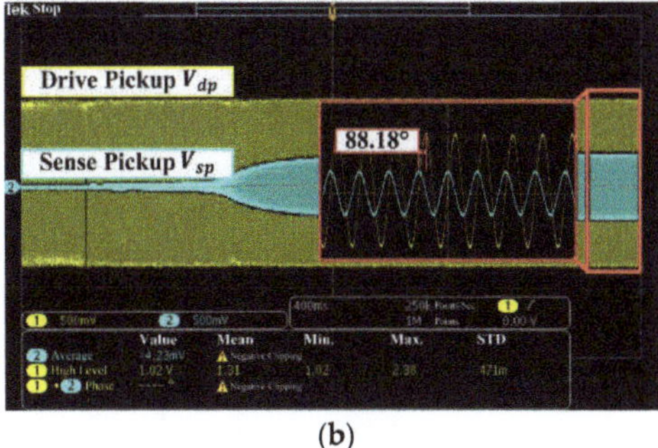

(b)

Figure 17. Start-up progress of (**a**) mode matching and (**b**) the sense mode.

The scale factors of both the mode matching and mismatching operations are tested as shown in Figure 18. The discrete points are the tested samples, and the solid lines are the corresponding first-order fitting lines within the range from $-50°/s$ to $40°/s$. The scale factor of the mode matching operation is calculated as 6.70 mV/°/s, which is improved 20.6 times compared with that of the mode mismatching operation (0.33 mV/°/s). This improvement in experimental scale factor conforms well to the estimated value in the corresponding system simulation. The nonlinearity of the mode matching operation is calculated as 1.77%, which is comparable to that of the mode mismatching operation of 1.61%. In addition, it is notable that the scale factor is deteriorated when Ω_{IN} exceeds $40°/s$, where Ω_{VC} is nearly offset by Ω_{IN} and cannot provide enough gain for the mode-matching loop. Herein, the mode mismatching occurs and decreases the scale factor.

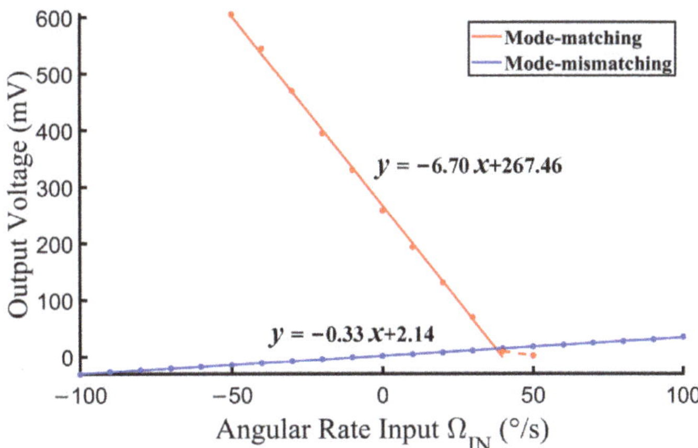

Figure 18. Scale factor test of mode matching and mismatching.

Finally, the Allan variances of both mode matching and mismatching operations are calculated as shown in Figure 19. As the sense mode of the gyro oscillates at its resonant frequency, the signal–noise ratio (SNR) is also improved. The data are acquired with the sampling rate of 10 Hz during 2 h. The bias instability (BI) of the mode matching operation is calculated as 9.60°/h, which is suppressed 3.25 times compared with that of the mode mismatching operation of 31.53°/h.

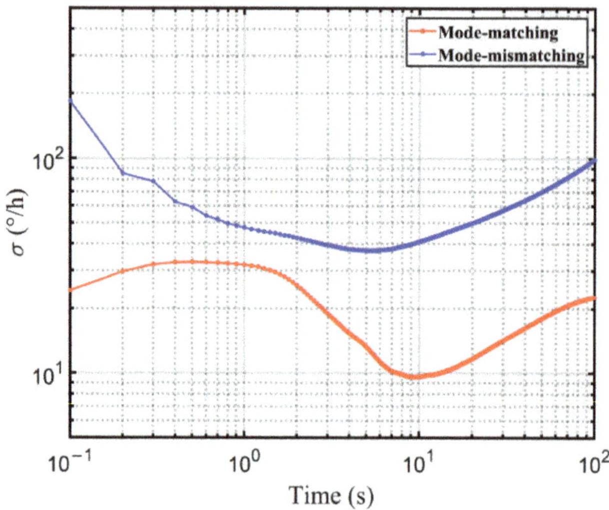

Figure 19. Allan variance test of mode matching and mismatching.

The comparison of the performance in different works is given in Table 4. The matching error of this work is at a similar level of [29,31], which indicates that the quadrature suppression circuit is unnecessary in a VCF-based mode-matching circuit relying on the proposed TFG decoupling design. For the specific numerical difference, the implementation of a prototype, such as the selection of operational amplifier or voltage reference, may be the primary influencing factor [29]. The BI suppression of this work is not outstanding for the lack of the test environment control. Notably, the work in [31] presents an extraordinary scale factor improvement and BI suppression. This feature probably benefits from a high-quality factor of 150,000 of the gyro.

Table 4. Summary of the performance of VCF-based works.

Ref.	Matching Error (Hz)	SF Improvement (Times)	BI Suppression (Times)
This work	0.014	20.6	3.28
[28]	0.6	5.60	-
[29]	0.03	16.8	-
[30]	0.1	17.9	10.9
[31]	0.005	68.8	23.1

5. Conclusions

In this paper, a simplified VCF-based mode-matching micromachine-optimized TFG is proposed. For the gyroscope sensor, the structure is optimized to provide a better performance of structural quadrature suppression. The related theoretical analyses, simulations, and experiments are implemented. The overall performance of the tested prototype is that the frequency split is narrowed from 20 Hz to 0.014 Hz. The scale factor is improved from 0.33 mV/°/s to 6.70 mV/°/s. The BI is suppressed from 31.53°/h to 9.60°/h. The level of BI agrees well with the accuracy requirement of the tactical grade and can satisfy applications, such as those in industrial electronics and those that are low-end tactical [1]. Benefiting from the simplicity of the circuit design, this gyroscope is particularly suitable in applications requiring a low cost and low power consumption, such as unmanned aerial vehicles (UAVs), movable energy harvesting systems, and consumer electronics. Alongside that, the tested matching error represents an outstanding degree of mode matching, and the times of BI suppression are at a similar level to that in other works. These results effectively prove the feasibility of the proposed non-quadrature-nulling mode matching scheme for avoiding the structure complexity, pick-up electrode occupation, and coupling between the quadrature-nulling loop and the mode-matching loop. As a preview of the work, it is hoped that it this will be applied in higher-end fields, such as attitude and heading reference systems (AHRSs) with a finer interface circuit to reduce the BI.

Author Contributions: Conceptualization, methodology, part of investigation, data curation, and writing-original draft preparation, Y.W.; supervision and project administration, W.Y.; part of investigation and part of draft editing, Y.X.; part of resources, H.C.; formal analysis, part of resources, writing—review, part of draft editing, and funding acquisition Q.S. All authors have read and agreed to the published version of the manuscript.

Funding: This research was funded in part by the Chinese National Science Foundation, grant number 52075454; in part by the Key Research and Development Program of Shaanxi Province, grant number 2021GY-283; and in part by the Aviation Science Foundation Project, grant number 201958053002.

Data Availability Statement: The data that support the findings of this study are available from the corresponding author upon reasonable request.

Conflicts of Interest: The authors declare no conflict of interest.

References

1. Passaro, V.; Cuccovillo, A.; Vaiani, L.; De Carlo, M.; Campanella, C. Gyroscope technology and applications: A review in the industrial perspective. *Sensors* **2017**, *17*, 2284. [CrossRef] [PubMed]
2. Liu, K.; Zhang, W.; Chen, W.; Li, K.; Dai, F.; Cui, F.; Wu, X.; Ma, G.; Xiao, Q. The development of micro-gyroscope technology. *J. Micromech. Microeng.* **2009**, *19*, 13001. [CrossRef]
3. Dell'Olio, F.; Brunetti, G.; Conteduca, D.; Sasanelli, N.; Ciminelli, C.; Armenise, M. Planar photonic gyroscopes for satellite attitude control. In Proceedings of the 2017 7th IEEE International Workshop on Advances in Sensors and Interfaces (IWASI), Venice, Italy, 15–16 June 2017; pp. 167–169.
4. Woodman, K.; Franks, P.; Richards, M. The nuclear magnetic resonance gyroscope: A review. *J. Navig.* **1987**, *40*, 366–384. [CrossRef]
5. Ezekiel, S.; Arditty, H. Fiber-optic rotation sensors. Tutorial review. In *Fiber-Optic Rotation Sensors and Related Technologies: Proceedings of the First International Conference MIT, Cambridge, MA, USA, 9–11 November 1981*; Springer: Berlin/Heidelberg, Germany, 1981; pp. 2–26.
6. Gadola, M.; Buffoli, A.; Sansa, M.; Berthelot, A.; Robert, P.; Langfelder, G. 1.3 mm^2 nav-grade NEMS-based gyroscope. *J. Microelectromech. Syst.* **2021**, *30*, 513–520. [CrossRef]
7. Wang, P.; Li, Q.; Zhang, Y.; Wu, Y.; Wu, X.; Xiao, D. Bias thermal stability improvement of mode-matching MEMS gyroscope using mode deflection. *J. Microelectromech. Syst.* **2023**, *32*, 1–3. [CrossRef]
8. Shen, Q.; Chang, H.; Wu, Y.; Xie, J. Turn-on bias behavior prediction for micromachined Coriolis vibratory gyroscopes. *Measurement* **2019**, *131*, 380–393. [CrossRef]
9. She, Q.; Yang, D.; Li, J.; Chang, H. Bias accuracy maintenance under unknown disturbances by multiple homogeneous MEMS gyroscopes fusion. *IEEE Trans. Ind. Electron.* **2023**, *70*, 3178–3187.
10. Shu, Y.; Hirai, Y.; Tsuchiya, T.; Tabata, O. Geometrical compensation of (100) single-crystal silicon mode-matched vibratory ring gyroscope. In Proceedings of the 2018 IEEE International Symposium on Inertial Sensors and Systems (INERTIAL), Lake Como, Italy, 26–29 March 2018.
11. Shu, Y.; Yoshikazu, H.; Toshiyuki, T.; Osamu, T. Geometrical compensation for mode-matching of a (100) silicon ring resonator for a vibratory gyroscope. *Jpn. J. Appl. Phys.* **2019**, *58*, SDDL06. [CrossRef]
12. Chen, J.; Tsukamoto, T.; Tanaka, S. Mode-Matched Multi-Ring Disk Resonator Using Single Crystal (100) Silicon. In Proceedings of the 2021 IEEE International Symposium on Inertial Sensors and Systems (INERTIAL), Kailua-Kona, HI, USA, 22–25 March 2021.
13. Wang, S.; Chen, J.; Tsukamoto, T.; Tanaka, S. Mode-matched multi-ring disk resonator using (100) single crystal silicon. In Proceedings of the 2022 IEEE 35th International Conference on Micro Electro Mechanical Systems Conference (MEMS), Tokyo, Japan, 9–13 January 2022.
14. Giner, J.; Maeda, D.; Ono, K.; Shkel, A.M.; Sekiguchi, T. MEMS gyroscope with concentrated springs suspensions demonstrating single digit frequency split and temperature robustness. *J. Microelectromechanical Syst.* **2019**, *28*, 25–35. [CrossRef]
15. Chen, J.; Tsukamoto, T.; Tanaka, S. Quad mass gyroscope with 16 ppm frequency mismatch trimmed by focus ion beam. In Proceedings of the 2019 IEEE International Symposium on Inertial Sensors and Systems (INERTIAL), Naples, FL, USA, 1–5 April 2019.

16. Schwartz, D.M.; Kim, D.; Stupar, P.; DeNatale, J.; M'Closkey, R.T. Modal parameter tuning of an axisymmetric resonator via mass perturbation. *J. Microelectromech. Syst.* **2015**, *24*, 545–555. [CrossRef]
17. Chen, C.; Wu, K.; Lu, K.; Li, Q.; Wang, C.; Wu, X.; Wang, B.; Xiao, D. A novel mechanical frequency tuning method based on mass-stiffness decoupling for MEMS gyroscopes. *Micromachines* **2022**, *7*, 1052. [CrossRef] [PubMed]
18. Li, C.; Wen, H.; Wisher, S.; Norouzpour-Shirazi, A.; Lei, J.Y.; Chen, H.; Ayazi, F. An FPGA-based interface system for high-frequency bulk-acoustic-wave microgyroscopes with in-run automatic mode-matching. *IEEE Trans. Instrum. Meas.* **2020**, *69*, 1783–1793. [CrossRef]
19. He, C.; Zhao, Q.; Huang, Q.; Liu, D.; Yang, Z.; Zhang, D.; Yan, G. A MEMS vibratory gyroscope with real-time mode-matching and robust control for the sense mode. *IEEE Sens. J.* **2015**, *15*, 2069–2077. [CrossRef]
20. Marx, M.; De Dorigo, D.; Nessler, S.; Rombach, S.; Manoli, Y. A 27 μW 0.06 mm^2 background resonance frequency tuning circuit based on noise observation for a 1.71 mW CT-ΔΣ MEMS gyroscope readout system with 0.9°/h bias instability. *IEEE J. Solid-State Circuits* **2018**, *53*, 174–186. [CrossRef]
21. Marx, M.; Cuignet, X.; Nessler, S.; De Dorigo, D.; Manoli, Y. An automatic mems gyroscope mode matching circuit based on noise observation. *IEEE Trans. Circuits Syst. II Express Briefs* **2019**, *66*, 743–747. [CrossRef]
22. Ding, X.; Ruan, Z.; Jia, J.; Huang, L.; Li, H.; Zhao, L. In-run mode-matching of mems gyroscopes based on power symmetry of readout signal in sense mode. *IEEE Sens. J.* **2021**, *21*, 23806–23817. [CrossRef]
23. Cheng, L.; Yang, B.; Guo, X.; Wu, L. A digital calibration technique of mems gyroscope for closed-loop mode-matching control. *Micromachines* **2019**, *10*, 496.
24. Sharma, A.; Zaman, M.F.; Ayazi, F. A sub-0.2°/hr bias drift micromechanical silicon gyroscope with automatic CMOS mode-matching. *IEEE J. Solid-State Circuits* **2009**, *44*, 1593–1608. [CrossRef]
25. Sonmezoglu, S.; Alper, S.E.; Akin, T. A high performance automatic mode-matched MEMS gyroscope with an improved thermal stability of the scale factor. In Proceedings of the 2013 Transducers & Eurosensors XXVII: The 17th International Conference on Solid-State Sensors, Actuators and Microsystems (TRANSDUCERS & EUROSENSORS XXVII), Barcelona, Spain, 16–20 June 2013.
26. Yesil, F.; Alper, S.E.; Akin, T. An automatic mode matching system for a high Q-factor MEMS gyroscope using a decoupled perturbation signal. In Proceedings of the Transducers—2015 18th International Conference on Solid-State Sensors, Actuators and Microsystems (TRANSDUCERS), Anchorage, AK, USA, 21–25 June 2015.
27. Xu, L.; Li, H.; Ni, Y.; Liu, J.; Huang, L. Frequency tuning of work modes in z-axis dual-mass silicon microgyroscope. *J. Sens.* **2014**, *2014*, 891735. [CrossRef]
28. Xu, L.; Li, H.; Yang, C.; Huang, L. Comparison of three automatic mode-matching methods for silicon micro-gyroscopes based on phase characteristic. *IEEE Sens. J.* **2015**, *16*, 610–619. [CrossRef]
29. Cao, H.; Zhang, Y.; Yan, J. Mode-matching control system design of silicon MEMS gyroscope. *J. Chin. Inert. Technol.* **2017**. [CrossRef]
30. Ruan, Z.; Ding, X.; Qin, Z.; Jia, J.; Li, H. Automatic mode-matching method for MEMS disk resonator gyroscopes based on virtual coriolis force. *Micromachines* **2020**, *11*, 210. [CrossRef] [PubMed]
31. Peng, Y.; Zhao, H.; Bu, F.; Yu, L.; Xu, D.; Guo, S. An automatically mode-matched MEMS gyroscope based on phase characteristics. In Proceedings of the 2019 IEEE 3rd Information Technology, Networking, Electronic and Automation Control Conference (ITNEC), Chengdu, China, 15–17 March 2019; pp. 2466–2470.
32. Wang, P.; Xu, Y.; Song, G.; Li, Q.; Wu, Y.; Wu, X.; Xiao, D. Calibration and compensation of the misalignment angle errors for the disk resonator gyroscopes. In Proceedings of the 2020 IEEE International Symposium on Inertial Sensors and Systems (INERTIAL), Hiroshima, Japan, 23–26 March 2020; pp. 1–3.
33. Wu, Z.; Feng, R.; Sun, C.; Wang, P.; Wu, G. A dual-mass fully decoupled MEMS gyroscope with optimized structural design for minimizing mechanical quadrature coupling. *Microelectron. Eng.* **2023**, *269*, 111918. [CrossRef]
34. Adams, S.G.; Bertsch, F.M.; Shaw, K.A.; Hartwell, P.G.; MacDonald, N.C.; Moon, F.C. Capacitance based tunable micromechanical resonators. In Proceedings of the International Conference on Solid-State Sensors and Actuators Conference, Stockholm, Sweden, 25–29 June 1995; pp. 438–441.
35. Hao, Y.; Xie, J.; Yuan, W.; Chang, H. Dicing-free SOI process based on wet release technology. *Micro Nano Lett.* **2016**, *11*, 775–778. [CrossRef]
36. Adams, S.G.; Bertsch, F.M.; Shaw, K.A.; MacDonald, N.C. Independent tuning of linear and nonlinear stiffness coefficients. *J. Microelectromech. Syst.* **1998**, *7*, 172–180. [CrossRef]

 *micromachines*

Article

Combined Temperature Compensation Method for Closed-Loop Microelectromechanical System Capacitive Accelerometer

Guowen Liu [1,2], Yu Liu [2], Zhaohan Li [2], Zhikang Ma [2], Xiao Ma [2], Xuefeng Wang [2], Xudong Zheng [1] and Zhonghe Jin [1,*]

[1] School of Aeronautics and Astronautics, Zhejiang University, Hangzhou 310058, China
[2] Beijing Institute of Aerospace Control Device, Beijing 100854, China
*   Correspondence: jinzh@zju.edu.cn

Abstract: This article describes a closed-loop detection MEMS accelerometer for acceleration measurement. This paper analyzes the working principle of MEMS accelerometers in detail and explains the relationship between the accelerometer zero bias, scale factor and voltage reference. Therefore, a combined compensation method is designed via reference voltage source compensation and terminal temperature compensation of the accelerometer, which comprehensively improves the performance over a wide temperature range of the accelerometer. The experiment results show that the initial range is reduced from 3679 ppm to 221 ppm with reference voltage source compensation, zero-bias stability of the accelerometer over temperature is increased by 14.3% on average and the scale factor stability over temperature is increased by 88.2% on average. After combined compensation, one accelerometer zero-bias stability over temperature was reduced to 40 μg and the scale factor stability over temperature was reduced to 16 ppm, the average value of the zero-bias stability over temperature was reduced from 1764 μg to 36 μg, the average value of the scale factor stability over temperature was reduced from 2270 ppm to 25 ppm, the average stability of the zero bias was increased by 97.96% and the average stability of the scale factor was increased by 98.90%.

Keywords: MEMS accelerometer; combined compensation; voltage reference; temperature compensation

Citation: Liu, G.; Liu, Y.; Li, Z.; Ma, Z.; Ma, X.; Wang, X.; Zheng, X.; Jin, Z. Combined Temperature Compensation Method for Closed-Loop Microelectromechanical System Capacitive Accelerometer. *Micromachines* **2023**, *14*, 1623. https://doi.org/10.3390/mi14081623

Academic Editor: Ion Stiharu

Received: 22 July 2023
Revised: 12 August 2023
Accepted: 15 August 2023
Published: 17 August 2023

1. Introduction

An accelerometer is a typical inertial sensor, which has a wide range of important applications in aviation, navigation, aerospace, weapons and civilian fields. However, the large size and high price of traditional accelerometers limit their application. With the development of MEMS (microelectromechanical system) technology, a variety of MEMS accelerometers have emerged, and their small size, small power consumption and wide application range have aroused the interest of research from all walks of life. At present, the sensitive structure of high-performance accelerometers mostly uses an all-silicon structure, which has developed rapidly due to its advantages of full-temperature performance, such as Safran's Colibrys sandwich all-silicon accelerometer with a zero-bias stability of 30 μg in 2020 [1]. As a typical representative, the Litton SiAC^{TM} silicon accelerometer has a range of more than 100 g, a zero bias better than 20 μg and a scale factor stability better than 50 ppm. In addition to the improvement of sensitive structures, there are some improvements in the circuit to enhance the accuracy of MEMS accelerometers. A 2012 Colibrys article introduced a navigation-grade Sigma-Delta MEMS accelerometer [2]. The accelerometer interface used a preamplifier and an ADC (analog-to-digital conversion) in part, and the rest of the circuitry was carried out digitally. At the same time, the closed-loop structure is adopted to reduce the equivalent noise and quantization noise of the structure, improve the linearity of the structure and ensure performance within the vibration environment.

Multiple ways have been proposed to improve the thermal behavior of MEMS accelerometers. Some studies propose the structure in [3–7], other studies reduce the thermal drift via compensating circuits and algorithm [8–19]. In 2015, Sergei A. Zotov et al. introduced a high-quality-factor resonant MEMS accelerometer [3]. To address drift over temperature, the MEMS sensor die incorporates two identical tuning forks with opposing axes of sensitivity. Demodulation of the differential FM output from the two simultaneously operated oscillators eliminates common mode errors and provides an FM output with continuous thermal compensation. Allan deviation of the differential FM accelerometer revealed a bias instability of 6 μg at 20 s, along with an elimination of any temperature drift due to increases in averaging time. In 2018, Giuseppe Ruzza et al. introduced the thermal compensation of low-cost MEMS accelerometers for tilt measurements [4], which have developed a miniaturized thermal chamber mounted on a tilting device to account for tilt angle variation. It can be determined whether it is warming or cooling the cycles, then select the corresponding compensation equation. After compensation, the RMS errors calculated for both the x- and y-axes decreased by 96%, but increased the complexity of the technique. In 2018, Wei Xu et al. reported a dual-differential accelerometer with an all-silicon structure with 3-times improved full temperature stability [5]. In 2020, Niu H et al. reported a comb accelerometer made of an all-silicon structure with a zero-bias stability of 100 μg [6]. In 2021, Liu Dandan et al. introduced an in situ compensation method for the scale factor temperature coefficient of a single-axis force-balanced MEMS accelerometer [7], which integrates the thermistor with the accelerometer to detect the temperature change of the accelerometer in real time, compensates for the change of the scale factor and reduces the temperature coefficient in the range of 25 °C to 50 °C for 6 ppm/°C.

In 2015, Qingjiang Wang et al. introduced the thermal characteristics of typical microelectromechanical system (MEMS) inertial measurement units (IMUs) with a reliable thermal test procedure [8]. The first-order piecewise function is introduced to establish the thermal models. The performance of both IMUs and inertial navigation systems improved significantly after compensation with the established thermal models. In 2019, Qing Lu et al. introduced a fusion algorithm-based temperature compensation method for a high-g MEMS accelerometer [10], which combines empirical mode decomposition (EMD), wavelet thresholding and temperature compensation to process measurement data from a high-g MEMS accelerometer. The experimental data show that the acceleration random walk changes from 1712.66 g/h/Hz$^{0.5}$ to 79.15 g/h/Hz$^{0.5}$ and the zero-deviation stability changes from 49,275 g/h to 774.7 g/h. In 2020, Vasco L et al. introduced a small-size, vacuum-packaged capacitive MEMS accelerometer through Sigma-Delta modulation [11], using an FPGA (field-programmable gate array) to achieve three different modulation levels, allowing for the flexible real-time adjustment of loop parameters. In 2021, Javier Martínez et al. introduced a lightweight thermal compensation technique for a MEMS capacitive accelerometer [12,13]. In this work, a light calibration method based on theoretical studies was proposed to obtain two characteristic parameters of the sensor's operation: the temperature drift of the bias and the temperature drift of the scale factor. This method requires less data to obtain the characteristic parameters, allowing for faster calibration. In 2021, Pengcheng Cai et al. introduced an improved difference temperature compensation method for MEMS resonant accelerometers [14], which proposed an improved temperature compensation approach, called proportional difference, for accelerometers based on differential frequency modulation. Experiment results demonstrate that the temperature sensitivity of the prototype sensor was reduced from 43.16 ppm/°C to 0.83 ppm/°C within the temperature range of −10 °C to 70 °C using the proposed method. In 2022, Qi Bing et al. introduced a novel accurate temperature drift error estimation model using microstructural thermal analysis [15], which obtains the complete temperature correlated quantities through structural thermal deformation. Moreover, the particle swarm optimization genetic algorithm back propagation neural network [16] was used to improve the accuracy and real-time recognition of accelerometer models. Compared with the traditional model, the accuracy was improved by 16%, and the number of iterations reduced by up to 99.86%. In 2023, Gangqiang Guo et al. introduced temperature drift compensation of a MEMS accelerometer based on DLSTM and ISSA [17], which improved bias instability, rate random walk and rate ramp with an increase of 96.68% on average. In 2023, Bo Yuan et al. presented a calibration and thermal compensation method for triaxial accelerometers based on the Levenberg–Marquardt (LM) algorithm and polynomial methods [18]. Within the temperature range of −40 °C to 60 °C, the temperature drifts of x- and y-axes reduced from −13.2 and 11.8 mg to −0.9 and −1.1 mg, respectively. The z-axis temperature drift was reduced from −17.9 to 1.8 mg. In 2023, Mingkang Li et al. reported an approach of in-operation temperature bias drift compensation based on phase-based calibration for a stiffness-tunable MEMS accelerometer with double-sided parallel plate (DSPP) capacitors [19]. The demodulated phase of the excited response exhibits a monotonic relationship with the effective stiffness of the accelerometer. Through the proposed online compensation approach, the temperature drift of the effective stiffness can be detected through the demodulated phase and compensated in real time by adjusting the stiffness-tuning voltage of DSPP capacitors. The temperature drift coefficient (TDC) of the accelerometer is reduced from 0.54 to 0.29 mg/°C, and the Allan variance bias instability of about 2.8 μg is not adversely affected.

Due to its excellent characteristics, all-silicon accelerometers can be used in navigation and guidance fields, such as the navigation and control of small UAVs, short-range tactical weapon guidance, etc. This article analyzes the operating principle of capacitive accelerometers and focuses on the factors that affect the full-temperature performance of capacitive accelerometers. In this paper, an all-silicon comb accelerometer with anchor zone stress cancellation technology will be analyzed [20], and a combined compensation method is designed to improve the full-temperature performance of the accelerometer via reference voltage source compensation and terminal temperature compensation of the accelerometer, which provides a basis for the development of capacitive accelerometers with a high range, high precision and high sensitivity.

2. MEMS Accelerometer Composition

As shown in Figure 1, the MEMS accelerometer includes and the accelerometer-sensitive structure which shown in Figure 2, capacitor–voltage conversion module, low-pass filter, PID controller, torque meter, temperature sensor, analog-to-digital conversion module and temperature compensation algorithm module [21,22].

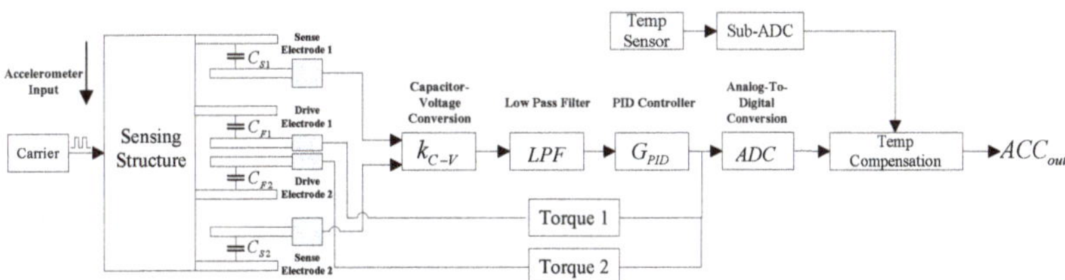

Figure 1. Block diagram of a MEMS accelerometer.

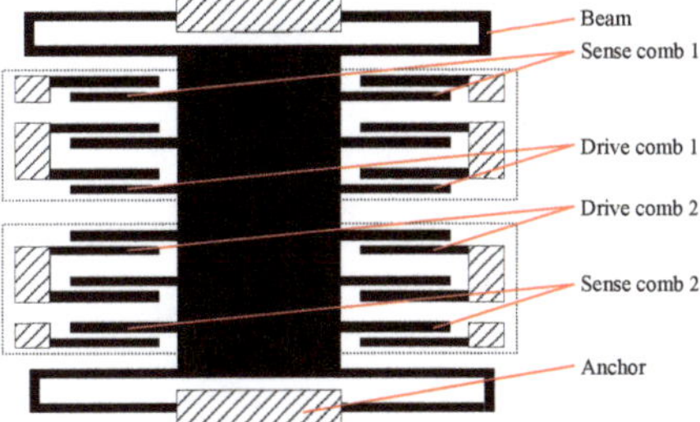

Figure 2. Schematic diagram of a MEMS accelerometer-sensitive structure.

3. MEMS Accelerometer Principle

As shown in Figure 1 above, when the external acceleration input occurs, the sensitive structure of the accelerometer is displaced relative to the carrier coordinate system. The distance between the sense electrode 1 and the sensing structure increases, so that the sense capacitance C_{S1} decreases, and the sense electrode 2 decreases with the sensitive structure comb electrode, so that the sense capacitance C_{S2} increases.

The sense capacitance C_{S1} and the sense capacitance C_{S2} convert the two capacitor values into two voltage values through the capacitance–voltage conversion circuit, and obtain the differential voltage via differential operation. This differential voltage is output to the analog-to-digital conversion module through a low-pass filter and PID controller. At the same time, the output of the PID controller is amplified by torque 1 and torque 2, and then feed back to the drive electrode 1 and 2, respectively. Drive electrode 1 forms drive capacitance with the sensing structure electrode C_{F1}, and drive electrode 2 forms drive capacitance with sensing structure electrode C_{F2}. Because the torque applied to the two drive voltages differs, the electrostatic attraction of C_{F1} is greater than that of C_{F2}, ultimately forcing the accelerometer-sensitive structure to remain near the initial position at all times. The output voltage of the PID controller is fed into the analog-to-digital conversion module and converted into a digital signal, and finally, the digital output of the temperature sensor performs temperature compensation calculations and outputs the final acceleration signal.

When the electrostatic force is equilibrated, the position of the movable comb (the electrical zero point) deviates from the mechanical zero point, δ. For the movable comb, the electrostatic force balance equation is:

$$K_m\delta + F_e + ma = 0 \tag{1}$$

In this equation, K_m is the mechanical stiffness of the accelerometer cantilever beam, F_e is the electrostatic force, m is the mass of the accelerometer sensitive structure and a is the acceleration input.

According to the static equilibrium analysis of the accelerometer closed-loop system [22], the output when the electrostatic force equilibrium is obtained is:

$$V_{fb} = \frac{V_R\left(\frac{d_0}{\delta} + \frac{\delta}{d_0}\right)}{2K_{fb}} \pm \frac{V_R\left(\frac{d_0}{\delta} - \frac{\delta}{d_0}\right)}{2K_{fb}}\sqrt{1 + \frac{2K_m}{C_f}\left(\frac{\delta}{V_R}\right)^2 + \frac{2ma\delta}{C_f V_R^2}} \tag{2}$$

Which can be expanded to:

$$V_{fb} = \frac{\delta V_R}{d_0 K_{fb}} - \frac{(K_m\delta + ma)d_0}{2C_f V_R K_{fb}} \tag{3}$$

In this equation, C_f is the sense capacitance, V_R is the driving reference voltage, K_m is the mechanical stiffness of the accelerometer cantilever beam, V_{fb} is the feedback output, K_{fb} is the feedback coefficient, m is the mass of the accelerometer sensitive structure, a is the acceleration input, d_0 is the comb gap and F_e is the electrostatic force.

Equation (3) provides a quadratic model of the static equilibrium state, so a zero-bias K_0 is:

$$K_0 = \frac{\delta V_R}{d_0 K_{fb}} - \frac{K_m\delta d_0}{2C_f V_R K_{fb}} \tag{4}$$

The primary term coefficient K_1 is:

$$K_1 = -\frac{md_0}{2C_f V_R K_{fb}} \tag{5}$$

As seen from Equations (4) and (5), the zero bias and scale factor are all related to the drive reference voltage V_R. From this, the temperature characteristics have a direct impact on the performance of the circuit. Figure 3 shows the flow direction of the main reference voltage of the closed-loop control circuit. It can be confirmed that the band-gap reference (BGR) used by the ASIC is the final source of the subsequent drive reference voltage, so its temperature characteristics have a direct impact on the temperature characteristics of the entire accelerometer.

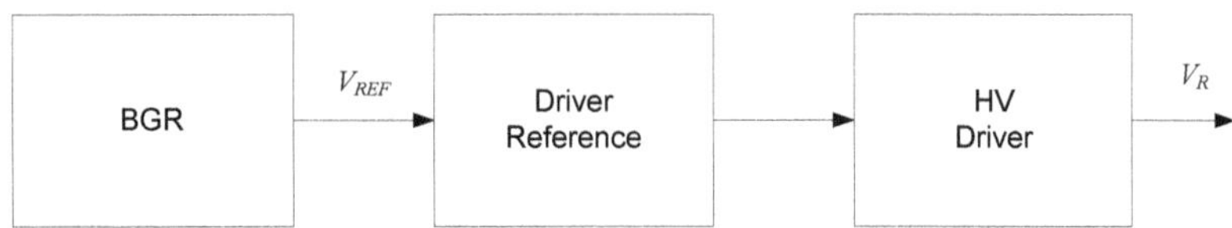

Figure 3. Accelerometer ASIC reference voltage source.

4. Temperature Compensation Method

4.1. Reference Voltage Source Compensation

A common band-gap reference circuit [23,24] is based on the principle of adding two voltages of equal magnitude and opposite temperature coefficients to obtain a temperature-independent voltage. The negative temperature coefficient voltage is realized through the base-emitter voltage, V_{BE}, of the substrate tertiary transistor. The positive temperature coefficient is accomplished through the base-emitter voltage difference, ΔV_{BE}, of two transistors operating at different current densities. In practice, the sum of the two temperature coefficients is not exactly zero. A typical band-gap reference is shown in the Figure 4 below.

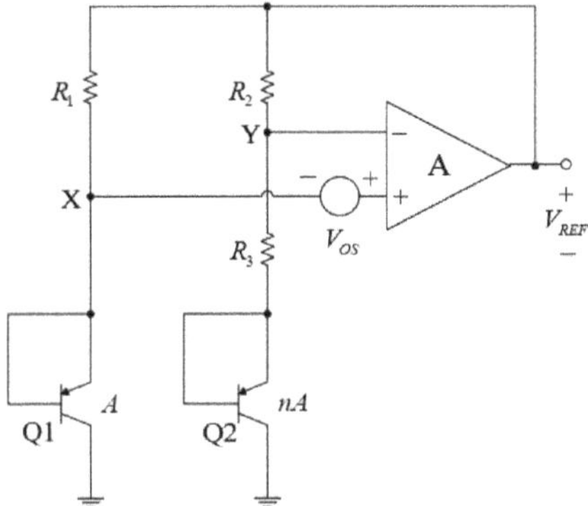

Figure 4. Band-gap reference source schematic.

In the Figure 4, V_{OS} is the offset voltage and the emitter area of bipolar transistor Q2 is n times that of Q1. Under the action of the op amp, the voltages of nodes X and Y are equal, the current flowing through bipolar transistors Q2 and Q1 is also equal and the base-emitter voltage difference between Q1 and Q2 is:

$$\Delta V_{BE} = V_T \ln n \tag{6}$$

$V_T = kT/q$, k is the Boltzmann constant, T is the absolute temperature and q is the electron charge. V_{BE2} is a negative temperature coefficient. The reference voltage output with the offset in this case is [23]:

$$V_{REF} = V_{BE2} + \left(1 + \frac{R_2}{R_3}\right)(V_T \ln n + V_{OS}) \tag{7}$$

By selecting the appropriate ratio of n to resistors R_2 and R_3, a temperature-independent output reference voltage can be obtained. In integrated circuit design, for symmetry considerations, n is generally taken as 8, and R_2 and R_3 are implemented through the resistor repair network and controlled via digital registers or other means.

The Figure 5 shows the simulation results of the temperature characteristics of the output voltage. The maximum change in output voltage is 1.55 mV and the temperature coefficient is 9 ppm/$°$C, from $-50\ °$C to $85\ °$C.

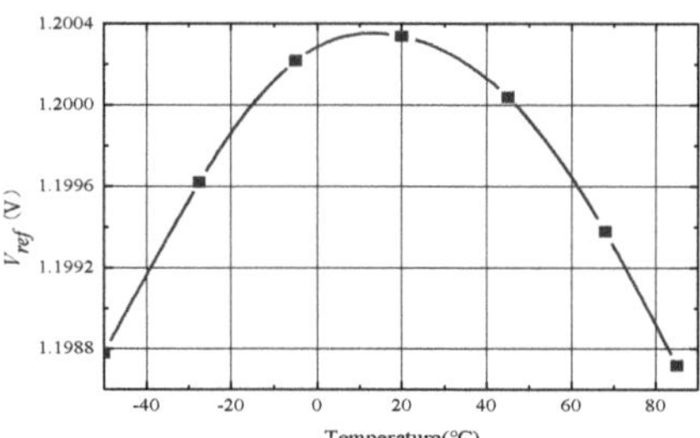

Figure 5. Simulation of temperature characteristics of band-gap references.

The number of transistors in the ASIC design process is a definite value. The temperature characteristics are determined by the acceleration sensor itself and are affected by the processing process. Thus, the resistor can be adjusted later by forming a resistor network.

According to Equation (5), the scale factor of the system is directly related to the high drive voltage, VR, so the temperature characteristics of the scale factor will be directly affected by the characteristics of the high drive voltage and therefore also by the temperature characteristics of the band-gap reference.

Limited by the ASIC chip pin count and anti-interference design, the direct output of the band-gap reference is not easy to measure, so we selected the 4.5 V output point of the chip to evaluate the temperature characteristics of the reference.

Due to the deviation between the ASIC design value and the actual processing, the actual measured temperature characteristics of the 4.5 V reference voltage are shown in Figure 6 below, with an initial range of 3679 ppm, and after adjusting the resistor network through multiple rounds of iterative experiments, the final range is 221 ppm, as shown in Figure 7.

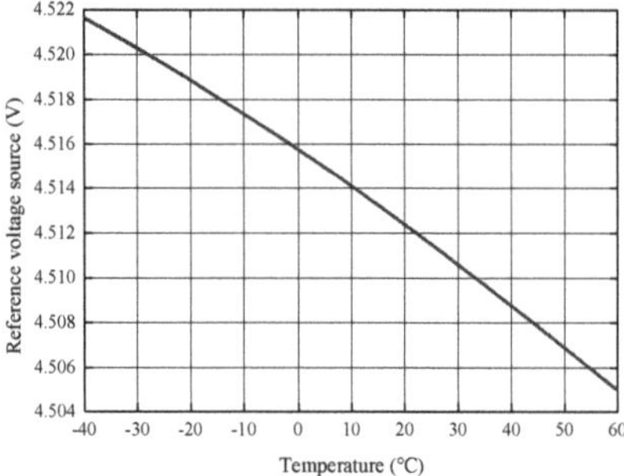

Figure 6. Factory state reference temperature characteristics of the circuit chip.

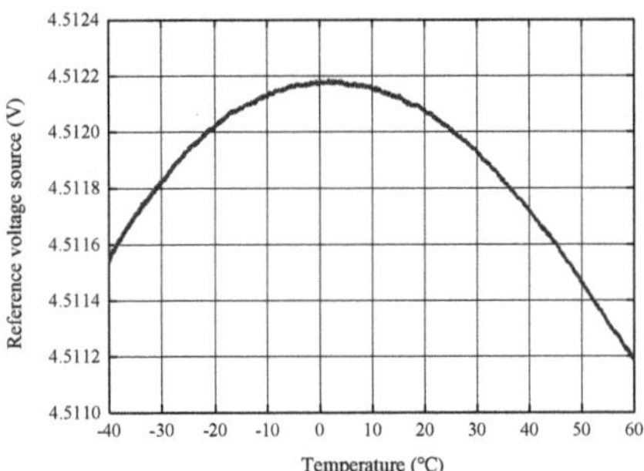

Figure 7. Modified reference temperature characteristics of the circuit chip.

4.2. Accelerometer Terminal Temperature Compensation

In order to further improve the performance over a wide temperature range of the accelerometer, the terminal third-order temperature compensation of the accelerometer is continued on the basis of the reference voltage source compensation to improve the performance over a wide temperature range of the accelerometer. Through the stability-over-temperature modeling experiment of the accelerometer, the temperature sensor output, T_i, zero-bias K_{0i} and scale factor, K_{1i}, of the accelerometer at each temperature point are obtained. The third-order polynomial model is fit to the zero bias vs. accelerometer temperature sensor output, where p_0, p_1, p_2 and p_3 are the 0–3-order coefficients of the model. Then, the accelerometer zero bias K_0 is:

$$K_0 = p_3 T^3 + p_2 T^2 + p_1 T + p_0 \tag{8}$$

The fitting error E is:

$$E = \sum_{i=0}^{n} \left[K_0 - \left(p_3 T_i^3 + p_3 T_i^2 + p_1 T_i + p_0 \right) \right]^2 \tag{9}$$

In order to minimize the fitting error, it is necessary to make its various deviations $\frac{\partial E}{\partial T_i} = 0$

$$-2\sum_{i=0}^{n} \left[K_0 - \left(+p_3 T_i^3 + p_3 T_i^2 + p_1 T_i + p_0 \right) \right] = 0$$

$$-2\sum_{i=0}^{n} \left[K_0 - \left(p_3 T_i^3 + p_3 T_i^2 + p_1 T_i + p_0 \right) \right] \cdot T_i^1 = 0 \tag{10}$$

$$-2\sum_{i=0}^{n} \left[K_0 - \left(p_3 T_i^3 + p_3 T_i^2 + p_1 T_i + p_0 \right) \right] \cdot T_i^2 = 0$$

$$-2\sum_{i=0}^{n} \left[K_0 - \left(p_3 T_i^3 + p_3 T_i^2 + p_1 T_i + p_0 \right) \right] \cdot T_i^3 = 0$$

Sorted out, it is:

$$
\begin{bmatrix}
n & \sum_{i=0}^{n} T_i & \sum_{i=0}^{n} T_i^2 & \sum_{i=0}^{n} T_i^3 \\
\sum_{i=0}^{n} T_i & \sum_{i=0}^{n} T_i^2 & \sum_{i=0}^{n} T_i^3 & \sum_{i=0}^{n} T_i^4 \\
\sum_{i=0}^{n} T_i^2 & \sum_{i=0}^{n} T_i^3 & \sum_{i=0}^{n} T_i^4 & \sum_{i=0}^{n} T_i^5 \\
\sum_{i=0}^{n} T_i^3 & \sum_{i=0}^{n} T_i^4 & \sum_{i=0}^{n} T_i^5 & \sum_{i=0}^{n} T_i^6
\end{bmatrix}
\cdot
\begin{bmatrix}
p_0 \\ p_1 \\ p_2 \\ p_3
\end{bmatrix}
=
\begin{bmatrix}
\sum_{i=0}^{n} K_{0i} \\
\sum_{i=0}^{n} K_{0i} T_i \\
\sum_{i=0}^{n} K_{0i} T_i^2 \\
\sum_{i=0}^{n} K_{0i} T_i^3
\end{bmatrix}
\tag{11}
$$

Let the van der Mond matrix V:

$$V = \begin{bmatrix} 1 & 1 & \cdots & 1 \\ T_1 & T_2 & \cdots & T_n \\ T_1^2 & T_2^2 & \cdots & T_n^2 \\ T_1^3 & T_2^3 & \cdots & T_n^3 \end{bmatrix} \tag{12}$$

Sorted out, it is:

$$VV^T = \begin{bmatrix} n & \sum\limits_{i=0}^{n} T_i & \sum\limits_{i=0}^{n} T_i^2 & \sum\limits_{i=0}^{n} T_i^3 \\ \sum\limits_{i=0}^{n} T_i & \sum\limits_{i=0}^{n} T_i^2 & \sum\limits_{i=0}^{n} T_i^3 & \sum\limits_{i=0}^{n} T_i^4 \\ \sum\limits_{i=0}^{n} T_i^2 & \sum\limits_{i=0}^{n} T_i^3 & \sum\limits_{i=0}^{n} T_i^4 & \sum\limits_{i=0}^{n} T_i^5 \\ \sum\limits_{i=0}^{n} T_i^3 & \sum\limits_{i=0}^{n} T_i^4 & \sum\limits_{i=0}^{n} T_i^5 & \sum\limits_{i=0}^{n} T_i^6 \end{bmatrix}, \ V \begin{bmatrix} K_{0_1} \\ K_{0_2} \\ \vdots \\ K_{0_n} \end{bmatrix} = \begin{bmatrix} \sum\limits_{i=0}^{n} K_{0_i} \\ \sum\limits_{i=0}^{n} K_{0_i} T_i \\ \sum\limits_{i=0}^{n} K_{0_i} T_i^2 \\ \sum\limits_{i=0}^{n} K_{0_i} T_i^3 \end{bmatrix} \tag{13}$$

Therefore:

$$VV^T \cdot \begin{bmatrix} p_0 \\ p_1 \\ p_2 \\ p_3 \end{bmatrix} = V \cdot \begin{bmatrix} K_{0_1} \\ K_{0_2} \\ \vdots \\ K_{0_n} \end{bmatrix} \tag{14}$$

Finally, the third-order fitting coefficient of the zero bias vs. accelerometer temperature sensor output is obtained:

$$\begin{bmatrix} p_0 \\ p_1 \\ p_2 \\ p_3 \end{bmatrix} = \left(VV^T\right)^{-1} V \cdot K_0 \tag{15}$$

Similarly, a third-order polynomial model of the scale factor change vs. the output of the accelerometer temperature sensor can be obtained:

$$\frac{K_1{}'}{K_{1_T}} = q_3 T^3 + q_2 T^2 + q_1 T + q_0 \tag{16}$$

$K_1{}'$ is the scale factor of the accelerometer at room temperature, K_{1_T} is the scale factor of the accelerometer at each temperature point and q_0, q_1, q_2 and q_3 are the 0–3-coefficients of the model.

The ASIC chip is known to read the temperature sensor data, T, and the accelerometer output measurement, D_a. According to Equations (8) and (16), the accelerometer temperature compensation value D_{out} is:

$$D_{out} = (D_a - K_0) \cdot \left(\frac{K_1{}'}{K_{1_T}} \right) \tag{17}$$

5. Comparative Experiments

5.1. Uncompensated Experiments

Temperature tests have been performed for the $-40\ ^\circ$C to $+60\ ^\circ$C range with a $1\ ^\circ$C/min temperature gradient. The test data collection is smoothed for 1 s, the bias and scale factor are calculated using the range and the stability is calculated using the standard deviation. Figure 8 shows the accelerometer product photo, where the mass is 0.7 g and size $9 \times 9 \times 2.7$ mm^3. Figure 9 shows the accelerometer temperature performance test system, model GWS EG-02JAS. The temperature uncompensated accelerometer's zero bias and scale factor as a function of temperature are shown in Figure 10 below. From Figure 10a, it can be seen that the change (peak–peak) of the zero bias of the five MEMS accelerometers with temperature compensation from −40 °C to +60 °C is distributed between 4098 μg and 7183 μg, and the zero-bias stability (1σ) over a wide temperature range is distributed between 1374 μg and 2400 μg. The result of each accelerometer varies linearly with temperature. From Figure 10b, it can be seen that the temperature variation (peak–peak) of the scale factor of the five MEMS accelerometers is distributed between 6135 ppm and 7347 ppm, and the stability of the scale factor over a wide temperature range (1σ) is distributed between 2075 ppm and 2472 ppm. The scale factor of each accelerometer is also basically linear against temperature.

Figure 8. Accelerometer product photo.

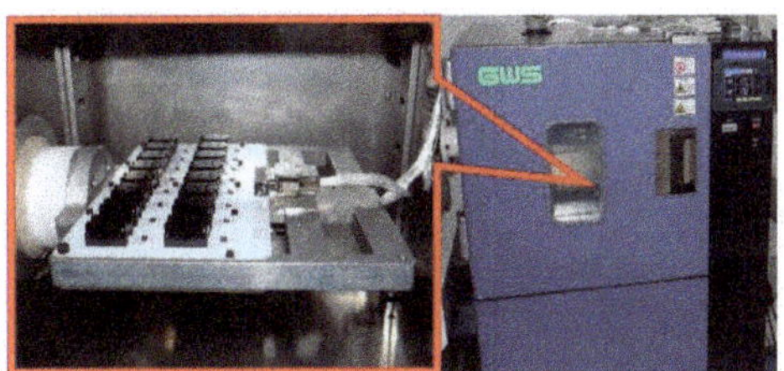

Figure 9. Accelerometer temperature performance test system.

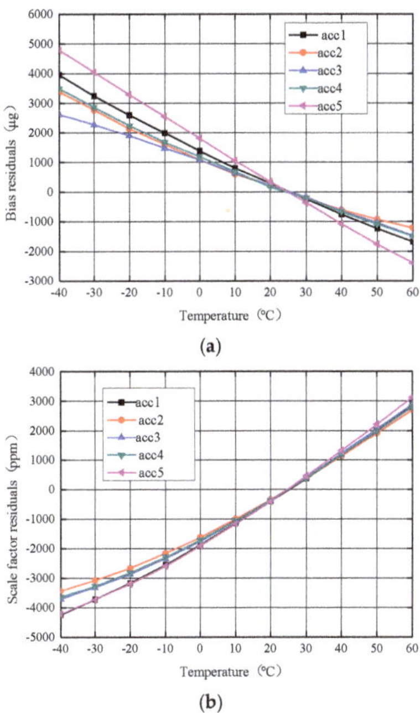

(a)

(b)

Figure 10. The reference voltage source is uncompensated for the zero bias and scale factor vs. temperature. (**a**) Zero bias curve against temperature. (**b**) Scale factor curve against temperature.

5.2. Accelerometer Stability over Temperature Experiment after Independent Reference Voltage Source Compensation

The zero bias and scale factor of the accelerometer with reference voltage source compensation as a function of temperature are shown in Figure 11. As seen in Figure 11a, after the reference voltage source compensation, the zero bias of the five MEMS accelerometers was distributed between 3367 µg and 6576 µg with the temperature from −40 °C to +60 °C, and the zero bias stability over temperature (1σ) was distributed between 1156 µg and 2196 µg, and the zero bias stability of each accelerometer increased by an average of 14.3%, with the result remaining linear against temperature. As seen from Figure 11b, the temperature variation (peak–peak) of the scale factor of the five MEMS accelerometers was distributed between 564 ppm and 1044 ppm, the stability of the scale factor over temperature (1σ) was distributed between 191 ppm and 330 ppm and the stability of scale factor of each accelerometer was increased by 88.2% on average. The comparative experiment results show that the accelerometer scale factor performance over temperature can be significantly improved by compensating the reference source temperature. The scale factor of each accelerometer as a function of temperature also changes from a basic linear curve to a second-order curve.

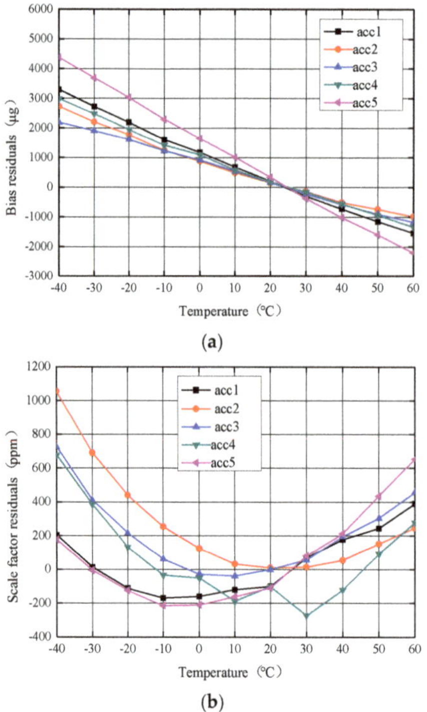

Figure 11. The zero bias and scale factor curves against temperature after reference voltage source compensation. (**a**) Zero bias curve against temperature. (**b**) Scale factor curve against temperature.

5.3. Accelerometer Stability over Temperature Experiment after Independent Terminal Temperature Compensation

To verify the effectiveness of the combined compensation method, the stability performance of the accelerometer after terminal temperature compensation was continued to be verified. Using the same five accelerometers and terminal temperature compensation, a plot of the accelerometer zero bias and scale factor as a function of temperature is obtained. Figure 12a shows that in the range of −40–+60 °C, the zero bias total temperature variation (peak–peak) of the five accelerometers is distributed in 83–317 µg, and the zero bias stability over temperature (1σ) is distributed in 26–85 µg. In the same temperature range, the total temperature variation (peak–peak) of the scale factor of the five accelerometers shown in Figure 12b is distributed in 71–175 ppm, and the stability of the scale factor over temperature (1σ) is distributed within 25–48 ppm. Among them, one accelerometer fluctuation is obviously more severe than the others, mainly because the processed accelerometer still has poor consistency and some degree of discreteness, but the overall results are still within the range. Terminal temperature compensation significantly improves accelerometer bias and scale factor performance over a wide temperature range.

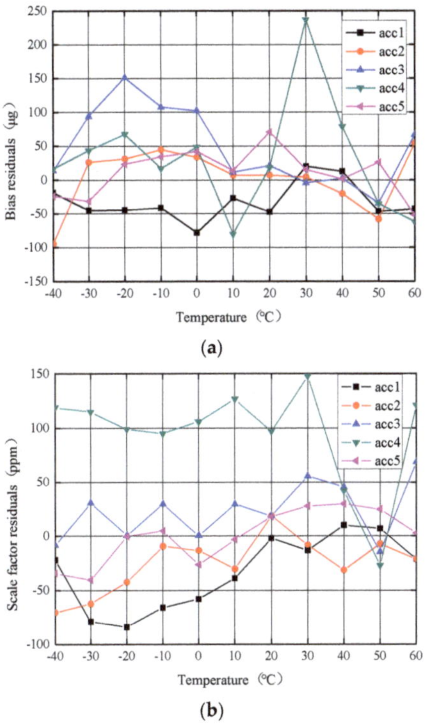

Figure 12. The zero bias and scale factor curve with temperature after terminal temperature compensation. (**a**) Zero bias curve against temperature. (**b**) Scale factor curve against temperature.

5.4. Stability over Temperature Experiment with Combined Temperature Compensation

The combined compensation of reference voltage source compensation and end-temperature compensation was used for the same five accelerometers to verify its validity. The zero bias and scale factor of the accelerometers as a function of temperature is shown in Figure 13. Figure 13a shows that in the range of −40–+60 °C, the zero bias temperature variation (peak–peak) of the five accelerometers was distributed within 79–183 µg, and the zero-bias stability over temperature (1σ) was distributed within 25–53 µg. Compared with only terminal temperature compensation, the combined compensation at the five accelerometers has improved the zero-bias stability at a wide temperature range by an average of 24.5%. Through combined temperature compensation, the total temperature variation (peak–peak) of the scale factor of the five accelerometers shown in Figure 13b was distributed within 43–145 ppm, and the stability of the scale factor over temperature (1σ) was distributed within 14–51 ppm. The scale factor stability of the five accelerometers was improved by an average of 25.5% compared to using only terminal temperature compensation. This combined compensation method comprehensively improves the accelerometer performance over a wide temperature range and proves the effectiveness of this method.

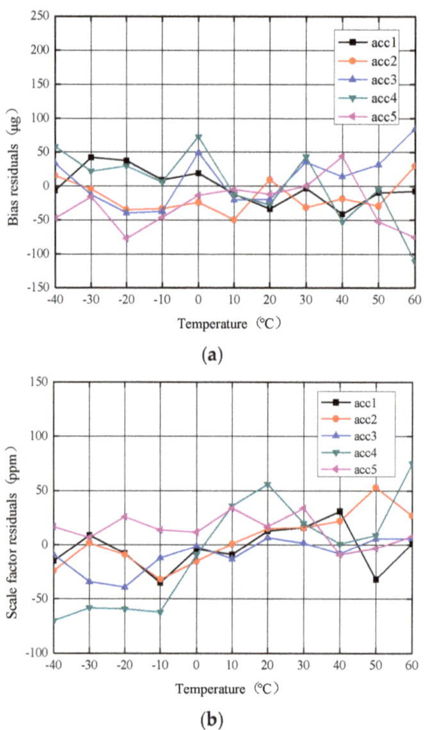

Figure 13. The zero bias and scale factor curve with temperature after the combined compensation. (**a**) Zero bias curve against temperature. (**b**) Scale factor curve against temperature.

Table 1 shows the zero bias and scale factor full temperature experiment data for an accelerometer numbered ACC3 before and after various compensations. Combined compensation has obvious advantages over any single compensation method.

Table 1. Comparison of data before and after various compensation for an ACC3 accelerometer.

Name	Uncompensated	Voltage Reference Source Compensation	Terminal Temperature Compensation	Combined Compensation
Zero bias variation over temperature (p–p) µg	4098	3367	185	124
Zero bias stability over temperature (1σ) µg	1374	1156	59	40
Scale factor variation over temperature (p–p) ppm	6606	765	83	46
Scale factor stability over temperature (1σ) ppm	2233	242	27	16

6. Conclusions

MEMS accelerometers are characterized by a small size, light weight and low cost. However, there is also a negative effect of the drift in temperature coefficient of the reference voltage source. This effect creates errors when the accelerometer operates at a wide temperature range, ultimately affecting the output signal of the accelerometer. In order to reduce the influence of the temperature coefficient of the reference voltage source, a method of temperature compensation of the reference voltage source is proposed, which is to reduce the influence of the temperature drift from the reference voltage source. To further improve the accelerometer performance over a wide temperature range, the accelerometer is compensated for the terminal third-order temperature compensation and the reference voltage source compensation. The experiment results show that the average value of zero-bias stability of the accelerometer with combined temperature compensation of reference voltage source compensation and terminal temperature compensation is reduced from 1764 μg to 36 μg, and the average stability of the scale factor over temperature is reduced from 2270 ppm to 25 ppm. The zero-bias stability over temperature is improved by 97.96% on average. The scale factor stability over temperature is improved by 98.90% on average. The combined compensation method greatly improves the accelerometer's performance over a wide temperature range, which increases the accelerometer's high-precision application capability. The accelerometer is suitable for environments with a short working time and minimal external temperature influence.

Author Contributions: Writing—original draft preparation, G.L., Y.L. and Z.L.; writing—review and editing, Y.L., Z.L., Z.M., X.M. and X.Z.; technical supervision, X.W. and Z.J. All authors have read and agreed to the published version of the manuscript.

Funding: The work was supported by the National Natural Science Foundation of China (no. 51976177) and Zhejiang Provincial Basic Public Welfare Research Program of China under grant no. LGG20F040001.

Data Availability Statement: Not applicable.

Conflicts of Interest: The authors declare no conflict of interest.

References

1. Marjoux, D.; Ullah, P.; Frantz-Rodriguez, N.; Morgado-Orsini, P.F.; Soursou, M.; Brisson, R.; Lenoir, Y.; Delhaye, F. Silicon MEMS By Safran—Navigation Grade Accelerometer Ready for Mass Production. In Proceedings of the 2020 DGON Inertial Sensors and Systems (ISS), Braunschweig, Germany, 15–16 September 2020; pp. 1–18. [CrossRef]
2. Zwahlen, P.; Dong, Y.; Nguyen, A.-M.; Rudolf, F.; Stauffer, J.-M.; Ullah, P.; Ragot, V. Breakthrough in High Performance Inertial Navigation Grade Sigma-Delta MEMS Accelerometer. In Proceedings of the 2012 IEEE/ION Position, Location and Navigation Symposium (PLANS), Myrtle Beach, SC, USA, 23–26 April 2012; pp. 15–19. [CrossRef]
3. Zotov, S.A.; Simon, B.R.; Trusov, A.A.; Shkel, A.M. High Quality Factor Resonant MEMS Accelerometer with Continuous Thermal Compensation. *IEEE Sens. J.* **2015**, *15*, 5045–5052. [CrossRef]
4. Ruzza, G.; Guerriero, L.; Revellino, P.; Guadagno, F.M. Thermal Compensation of Low-cost MEMS Accelerometers for Tilt Measurements. *Sensors* **2018**, *18*, 2536. [CrossRef] [PubMed]
5. Xu, W.; Tang, B.; Xie, G.; Yang, J. An All-Silicon Double Differential MEMS Accelerometer with Improved Thermal Stability. In Proceedings of the 2018 IEEE SENSORS, New Delhi, India, 28–31 October 2018; pp. 1–4. [CrossRef]
6. Niu, H.; Sun, G.; Wang, S.; Zhang, F. A Design of Capacitance MEMS Accelerometer with Wafer Level Encapsulated All-Silicon Comb Tooth. *J. Chin. Inert. Technol.* **2020**, *28*, 672–676. [CrossRef]
7. Liu, D.; Wu, W.; Yan, S.; Xu, Q.; Wang, Y.; Liu, H.; Tu, L. In-Situ Compensation on Temperature Coefficient of The Scale Factor for A Single-Axis Nano-G Force-Balance MEMS Accelerometer. *IEEE Sens. J.* **2021**, *21*, 19872–19880. [CrossRef]
8. Wang, Q.; Li, Y.; Niu, X. Thermal Calibration Procedure and Thermal Characterization of Low-cost Inertial Measurement Units. *J. Navig.* **2015**, *69*, 373–390. [CrossRef]
9. Gheorghe, M. Advanced Calibration Method, with Thermal Compensation, for 3-Axis MEMS Accelerometers. *Rom. J. Inf. Sci. Technol.* **2016**, *19*, 255–268.
10. Lu, Q.; Shen, C.; Cao, H.; Shi, Y.; Liu, J. Fusion Algorithm-Based Temperature Compensation Method for High-G MEMS Accelerometer. *Shock. Vib.* **2019**, *2019*, 3154845. [CrossRef]
11. Lima, V.; Cabral, J.; Kuhlmann, B.; Gaspar, J.; Rocha, L. Small-Size MEMS Accelerometer Encapsulated in Vacuum Using Sigma-Delta Modulation. In Proceedings of the 7th IEEE International Symposium on Inertial Sensors & Systems, Hiroshima, Japan, 23–26 March 2020. [CrossRef]
12. Martínez, J.; Asiain, D.; Beltrán, J.R. Lightweight Thermal Compensation Technique for MEMS Capacitive Accelerometer Oriented to Quasi-Static Measurements. *Sensors* **2021**, *21*, 3117. [CrossRef] [PubMed]
13. Martínez, J.; Asiain, D.; Beltrán, J.R. Factory Oriented Technique for Thermal Compensation in MEMS Capacitive Accelerometers. *Eng. Proc.* **2021**, *10*, 4. [CrossRef]

14. Cai, P.; Xiong, X.; Wang, K.; Wang, J.; Zou, X. An Improved Difference Temperature Compensation Method for MEMS Resonant Accelerometers. *Micromachines* **2021**, *12*, 1022. [CrossRef] [PubMed]
15. Qi, B.; Shi, S.; Zhao, L.; Cheng, J. A Novel Temperature Drift Error Precise Estimation Model for MEMS Accelerometers Using Microstructure Thermal Analysis. *Micromachines* **2022**, *13*, 835. [CrossRef] [PubMed]
16. Wang, S.; Zhu, W.; Shen, Y.; Ren, J.; Gu, H.; Wei, X. Temperature compensation for MEMS resonant accelerometer based on genetic algorithm optimization backpropagation neural network. *Sens. Actuators A Phys.* **2020**, *316*, 112393. [CrossRef]
17. Guo, G.; Chai, B.; Cheng, R.; Wang, Y. Temperature Drift Compensation of a MEMS Accelerometer Based on DLSTM and ISSA. *Sensors* **2023**, *23*, 1809. [CrossRef] [PubMed]
18. Yuan, B.; Tang, Z.; Zhang, P.; Lv, F. Thermal Calibration of Triaxial Accelerometer for Tilt Measurement. *Sensors* **2023**, *23*, 2105. [CrossRef] [PubMed]
19. Li, M.; Ma, Z.; Zhang, T.; Jin, Y.; Ye, Z.; Zheng, X.; Jin, Z. Temperature Bias Drift Phase-Based Compensation for a MEMS Accelerometer with Stiffness-Tuning Double-Sided Parallel Plate Capacitors. *Nanomanufacturing Metrol.* **2023**, *6*, 22. [CrossRef]
20. Liu, G.; Liu, Y.; Ma, X.; Wang, X.; Zheng, X.; Jin, Z. Research on a Method to Improve the Temperature Performance of an All-Silicon Accelerometer. *Micromachines* **2023**, *14*, 869. [CrossRef] [PubMed]
21. Chen, W.; Wang, T.; Yin, L.; Chen, X.; Guo, Y.; Liu, X. Closed-loop system of a capacitive micro-accelerometer. *J. Harbin Inst. Technol.* **2010**, *42*, 1720–1723.
22. Wu, T.; Dong, J.; Liu, Y. Closed-Loop System Bias in A Capacitive Micro-Accelerometer. *J. Tsinghua Univ. (Sci. Tech.)* **2005**, *45*, 201–204.
23. Hilbiber, D.F. A new semiconductor voltage standard. In Proceedings of the IEEE International Conference on Solid-State Circuits (ISSCC), Philadelphia, PA, USA, 19–21 February 1964; pp. 32–33. [CrossRef]
24. Kuijk, K.E. A precision reference voltage source. *IEEE J. Solid-Circuits* **1973**, *8*, 222–226. [CrossRef]

micromachines

Article

Research on a Method to Improve the Temperature Performance of an All-Silicon Accelerometer

Guowen Liu [1,2], Yu Liu [2], Xiao Ma [2], Xuefeng Wang [2], Xudong Zheng [1] and Zhonghe Jin [1,*]

[1] School of Aeronautics and Astronautics, Zhejiang University, Hangzhou 310058, China
[2] Beijing Institute of Aerospace Control Device, Beijing 100854, China
* Correspondence: jinzh@zju.edu.cn

Abstract: This paper presents a novel method for the performance of an all-silicon accelerometer by adjusting the ratio of the Si-SiO$_2$ bonding area, and the Au-Si bonding area in the anchor zone, with the aim of eliminating stress in the anchor region. The study includes the development of an accelerometer model and simulation analysis which demonstrates the stress maps of the accelerometer under different anchor–area ratios, which have a strong impact on the performance of the accelerometer. In practical applications, the deformation of the comb structure fixed by the anchor zone is influenced by the stress in the anchor region, causing a distorted nonlinear response signal. The simulation results demonstrate that when the area ratio of the Si-SiO$_2$ anchor zone to the Au-Si anchor zone decreases to 0.5, the stress in the anchor zone decreases significantly. Experimental results reveal that the full-temperature stability of zero-bias is optimized from 133 µg to 46 µg when the anchor–zone ratio of the accelerometer decreases from 0.8 to 0.5. At the same time, the full-temperature stability of the scale factor is optimized from 87 ppm to 32 ppm. Furthermore, zero-bias full-temperature stability and scale factor full-temperature stability are improved by 34.6% and 36.8%, respectively.

Keywords: MEMS accelerometer; response signal; anchor zone; stress cancellation

1. Introduction

Accelerometers are widely used in various areas, including navigation, aviation, aerospace, weapons, and civil fields. However, the traditional accelerometer's large size and high cost limits its applications. With the development of microelectromechanical systems (MEMS) technology, various MEMS accelerometers have emerged. Its characteristics of small size, low-power consumption, and wide application range have aroused the research interest of all circles [1]. N. Yazdi introduced an all-silicon structure accelerometer in approximately 2000 [2]. All-silicon structure accelerometers have the advantage of low temperature sensitivity and good long-term stability due to their material consistency. Colibrys, a division of Safran, introduced a new structure of the all-silicon sandwich accelerometer in 2020 [3], which improved the zero-bias full temperature stability to 30 µg. In 2016, D. Xiao reported a dual-differential torsional MEMS glass–silicon–glass sandwich accelerometer structure, in which the characteristics of the temperature coefficient were five times lower than before [4]. In 2018, Wei Xu reported an all-silicon structure of the dual-differential accelerometer, in which the total temperature stability was increased by three times [5]. In 2020, H. Niu reported that the same type of accelerometers had a zero-bias stability of 100 µg [6]. In 2018, Huan Liu reported two versions of capacitive accelerometers based on low-temperature co-fired ceramic (LTCC) technology, presenting a larger full-scale range (10 g), and lower nonlinearity of less than 1%, as well as a sensitivity of 30.27 mV/g [7]. The MEMS Technology Center at the Middle East University of Science and Technology in Turkey has manufactured a Silicon-On-Insulator (SOI) structure of the triaxial capacitive accelerometer, which had a background noise of 8 µg/$\sqrt{\text{Hz}}$, as recorded in 2020 [8]. In 2021, Yurong He reported a novel teeter-totter type accelerometer based on

Citation: Liu, G.; Liu, Y.; Ma, X.; Wang, X.; Zheng, X.; Jin, Z. Research on a Method to Improve the Temperature Performance of an All-Silicon Accelerometer. *Micromachines* **2023**, *14*, 869. https://doi.org/10.3390/mi14040869

Academic Editor: Ion Stiharu

Received: 24 March 2023
Revised: 14 April 2023
Accepted: 16 April 2023
Published: 18 April 2023

glass–silicon composite wafers, in which a zero-bias stability under 0.2 mg, and a noise floor with 11.28 µg/$\sqrt{\text{Hz}}$ were obtained [9]. In 2022, Yongjun Zhou reported an improved variational mode decomposition (VMD), and the time-frequency peak filtering (TFPF) denoising method has a smaller amount of signal distortion and a stronger denoising ability, so it can be adopted to denoise the output signal of the High-G MEMS accelerometer in order to improve its accuracy [10]. Litton SiAC^{TM} has reported a silicon-based MEMS accelerometer made of an all-silicon structure, with a measurement range of over 100 g, it has good characteristics including zero-bias stability and a scale factor stability that is better than 20 µg and 50 ppm.

The working mechanism of the accelerometer is based on Newton's law of inertia. It is a mechanical sensitive device, so all kinds of stresses will bring the output error of the accelerometer and deteriorate the full-temperature performance of the accelerometer. At present, there are three main measures to reduce the stress effect for accelerometers. First, stress isolation can be achieved by adding a stress isolation frame and stress isolation beam, or by low-stress encapsulation [11–13]. Second, the effect of in-plane stress is reduced by stress difference using differential symmetry structures [14]. Third, an all-silicon wafer level packaging process can be used to reduce the thermal stress caused by inconsistent coefficients of thermal expansion between layers [15]. The all-silicon wafer level packaging process can reduce the thermal stress caused by the inconsistent thermal expansion coefficients of the cap layer, the sensitive structure layer, and the substrate layer. However, the stress isolation structure will increase the acceleration sensor volume. Although the above low-stress packaging technology can reduce the external stress interference, it cannot reduce the internal stress of the chip. The differential symmetric structure mainly reduces the stress by the symmetric structure in X, Y, and Z directions. In this paper, based on the above stress-reduction technology, the internal stress reduction is further studied to adapt to the complex multi-layer structure of the MEMS accelerometer. Therefore, a bonding anchor zone stress cancellation method is proposed to reduce thermal stress transferred to the sensitive capacitor, which improves the accelerometer's overall temperature performance. This study provides a theoretical basis for the development of high-precision capacitive MEMS accelerometers.

2. MEMS Accelerometer Layer Design

Figure 1 shows the chip layer design of an all-silicon accelerometer, which consists of three layers: substrate layer, sensitive structure layer, and cap layer. A silicon dioxide graphic layer is present between the cap layer and the structure layer. The structure layer is bonded to the cap layer using Si-SiO$_2$ bonding, while the gold-silicon eutectic bonding is used to bond the sensitive structure layer to the substrate layer. This bonding arrangement creates a cavity in which the micro-structure can move freely. Electrodes are patterned on the substrate layer, and a coplanar electrode is used to connect the sensitive structure inside the cavity with the outer electrode pad. The anchor area is the bonding area between the cap layer and the sensitive structure layer, connecting the three layers together and supports the movable sensitive structure.

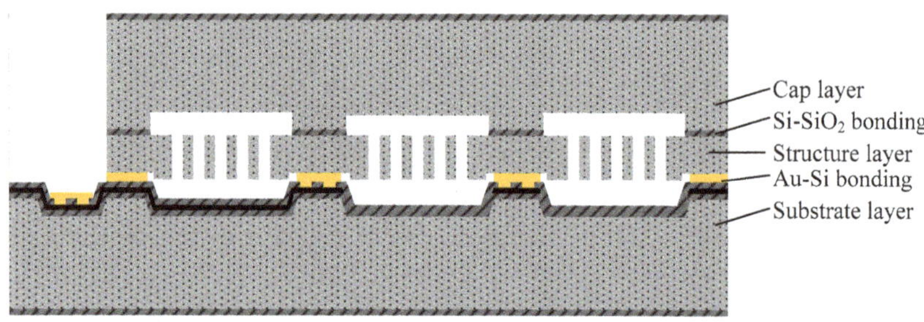

Figure 1. Structural Diagram of MEMS Accelerometer.

3. Principle of MEMS Accelerometer

As shown in Figure 2, the MEMS accelerometer system consists of several components, including accelerometer-sensitive structure, capacitance-voltage conversion module, low-pass filter (LPF), proportional integral derivative (PID) controller, torquer, analog-to-digital conversion module (ADC), digital bandwidth filtering module, and digital average filtering module. Under external input acceleration, the accelerometer sensitive structure moves relative to the carrier coordinate system. The distance between the sense electrode 1 and the movable comb electrode increases, so the sensing capacitance C_{S1} decreases, while the distance between the sense electrode 2 and the movable comb electrode decreases, the sensing capacitance C_{S2} increases. The sensing capacitance C_{S1} and the sensing capacitance C_{S2} are converted into two voltage values through a capacitor-voltage conversion circuit (CV), and then converted to differential voltage, which is sampled by the analog-to-digital converter module after passing through a low-pass filter and PID controller. After that, it is then passed through a bandpass filter and digital average filter to work as the output signal of the accelerometer. The output of the PID controller is amplified by the torque 1 and torque 2 circuits and fed back to the drive electrode 1 and drive electrode 2 of the accelerometer, respectively. The drive electrode 1 and the movable comb electrode form the drive capacitor C_{F1}, and the drive electrode 2 and the movable comb electrode form the drive capacitor C_{F2}. The electrostatic attraction of C_{F1} is greater than that of C_{F2} due to the different voltages applied by the torques to the two drive capacitors, which keeps the accelerometer-sensitive structure static near the initial position and thus forms an electrostatic balance closed-loop system.

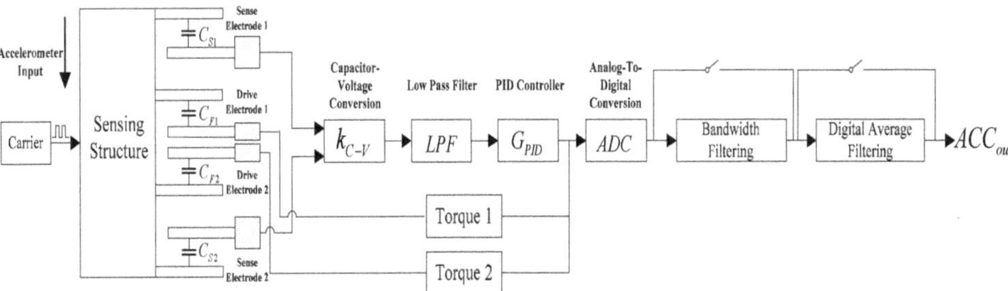

Figure 2. Principle block diagram of MEMS accelerometer.

4. Structural Stress Simulation Analysis of MEMS Accelerometer

The comb-tooth capacitive MEMS accelerometer described in this paper has a symmetrical arrangement of its sensitive structure layer. However, the cap layer and substrate layer above and below the structure layer are not completely symmetric. The anchor layer between the cap layer and the structure layer is made of silicon dioxide, while the anchor layer supported by the substrate layer and the sensitive structure layer is made of gold. It can be seen from the material properties in Table 1 that the thermal expansion coefficients of these materials are inconsistent, causing different stresses on the anchor area on both sides of the structure at different temperatures, which reduces the accelerometer's temperature performance. To address this issue, a stress cancellation method in the anchor zone is proposed in this paper. The optimal stress of the sensitive structure over full-temperature can be achieved by matching the anchor area of the upper and lower sides of the sensitive structure to achieve the balance of stress on both sides. The asymmetric stress is reduced drastically, and thus the full-temperature accuracy of the MEMS accelerometer is improved.

Table 1. Accelerometer Layer Material Attribute Table.

Name	Silicon	Silicon Dioxide	Au	Unit
Coefficient of thermal expansion (CTE)	2.60×10^{-6}	0.5×10^{-6}	14.2×10^{-6}	1/K
Constant pressure heat capacity	700	730	129	J/(kg·K)
Density	2329	2200	19,300	kg/m^3
Thermal conductivity	130	1.4	317	W/(m·K)
Young's modulus	1.69×10^{11}	7.00×10^{10}	7.00×10^{10}	Pa
Poisson's ratio	0.28	0.17	0.44	1

The stress generated by the bonding of the MEMS accelerometer is related to the anchor zone, and the balance of the stress on both sides of the structure layer is achieved by designing the size of the anchor zone on both sides of the sensitive structure layer, so that the stress of the sensitive structure is optimized at full-temperature, and the generation of asymmetric stress is fundamentally reduced, thereby improving the full-temperature accuracy of the accelerometer. Since the normal force and the lateral force are related to Poisson's ratio of the material, the lateral force has a strong correlation with the normal force, so optimizing the normal force also indirectly optimizes the lateral force, which may obtain a similar result in this research. Therefore, only the normal force is considered as the research object in this paper. The positive stress is multiplied by the area of the anchor area in order to obtain the stress value, and the stress on both sides of the anchor zone is treated with a combined force, and the point with the smallest resultant force is the most advantageous. By simulating the force curves of different anchor–zone ratios, the optimized result is selected and tested. The design of the acceleration is only different in the anchor zone, and the processing is consistent, so as to minimize external interference factors as much as possible. Finally, the change in the output result is strongly correlated with the change in the anchor–zone ratios, in order to complete the experimental verification.

Since the in-plane structure is fully symmetric, each anchor area has the same size. During simulation, a group of anchor areas are taken for modeling, as shown in Figure 3. The model consists of 10 layers, and the upper and lower layers of the structure are the silicon dioxide anchor layer and the gold silicon anchor layer, respectively.

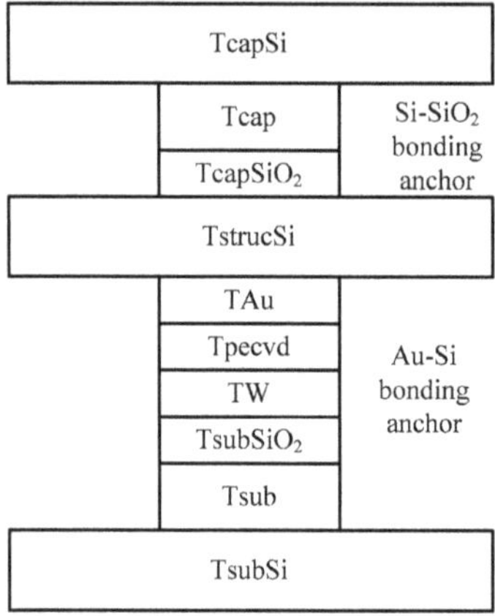

Figure 3. Anchorage simulation modeling model.

Figure 4 shows the simulation diagram of the stress model of the anchor zone, where the stress difference of the anchor zone is the added stress difference of the Si-SiO$_2$ zone and the Au-Si zone at different temperatures. The force on the anchor zone is the surface normal stress multiplied by the area of the anchor zone. The anchor–zone ratio is defined as the square root of the ratio of the area of the Si-SiO$_2$ anchor zone divided by the area of the Au-Si anchor zone.

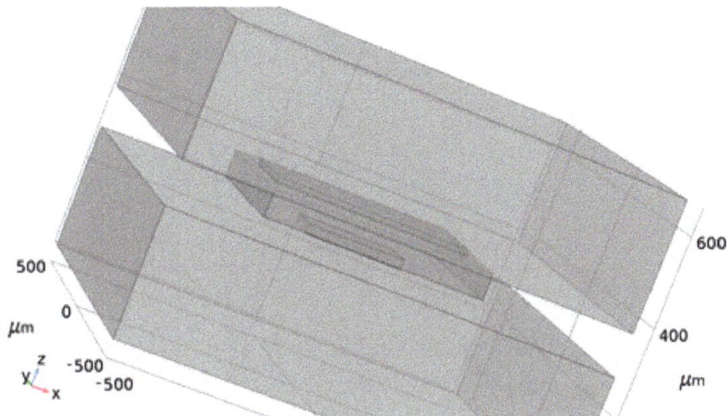

Figure 4. Anchor zone simulation diagram.

With the size of the silicon oxide anchor area of the current MEMS accelerometer fixed, the anchor–zone ratio varies from 0.1 to 2, and the simulations are carried out under the temperature conditions of 233.15 K, 278.15 K and 333.15 K, respectively. The simulation results are shown in Figure 5, which is the relationship between the ratio of the anchor zone and the force difference at three typical temperatures. The undulation of the curve is caused by the nonlinear nature of the material in each layer. As can be seen from the figure, the variation trends of the three curves are basically the same, and the regions with anchor–zone ratios of less than 0.6 all have gentle changes. Moreover, all three curves have the best force difference when the anchor–zone ratio is less than 0.6. The minimum force difference at the temperature 233.15 K is 5.33×10^{-4} N with the anchor–zone ratio of 0.1, while at temperatures of 278.15 K and 333.15 K, the smallest force difference is 7.49×10^{-5} N and 8.31×10^{-6} N, respectively, with the same anchor–zone ratio of 0.3. The simulation results are presented such that the force difference of the accelerometer-sensitive structure has a great consistency with the change trend of the anchor region ratio at different temperatures. The force of the accelerometer-sensitive structure at different temperatures can be reduced by selecting an appropriate anchor–zone ratio, so as to improve the temperature performance of the accelerometer.

Taking the risks involved in actual MEMS fabricating into consideration, a small anchor area means small bonding strength. To ensure the rationality of the design and process of the MEMS accelerometer, the anchor–zone ratio of 0.1 and 0.3 is considered poor. Thus, two ratios of 0.5 and 1.5 are chosen according to the force difference diagram as shown above for different anchor–zone ratios. Therefore, the force difference fluctuation of the anchor region ratio near 1.5 is too large, and it is easy to cause a large difference because of a micromachining error. Thus, the anchor–zone ratio of 0.5 is selected as the optimal value, which ensures that the structural force is relatively small within a certain process error range. Figures 6–8 show the maximum normal stress diagram of the Au-Si anchor zone at 233.15 K, 278.15 K and 333.15 K, respectively, with a 0.5 anchor–zone ratio. Figures 6–8 show the maximum normal stress diagram of the Au-Si anchor zone at 233.15 K, 278.15 K and 333.15 K, respectively, with a 0.5 anchor–zone ratio.

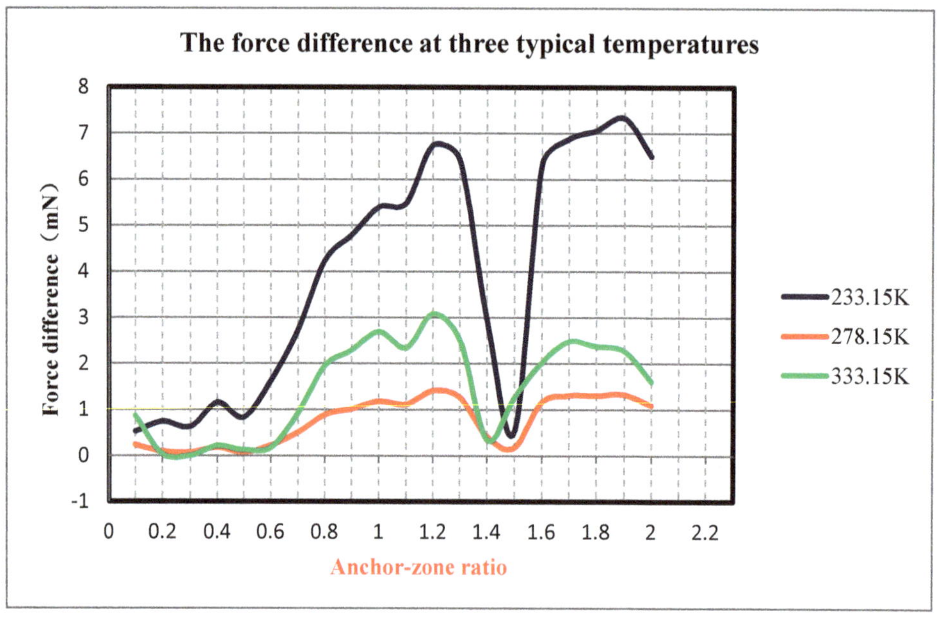

Figure 5. Diagram of anchor zone force difference under different anchorage zone ratios.

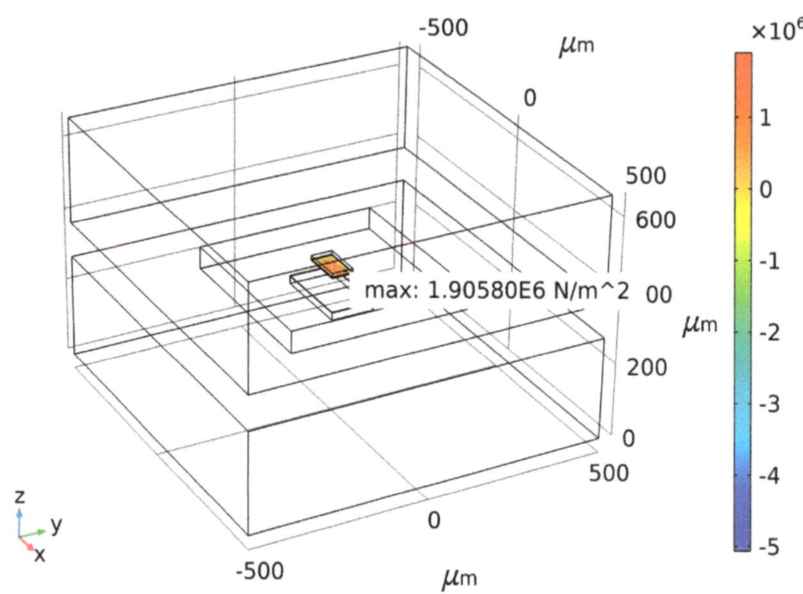

Figure 6. The maximum normal stress diagram of Au-Si anchor zone at 233.15 K with 0.5 anchor–zone ratio.

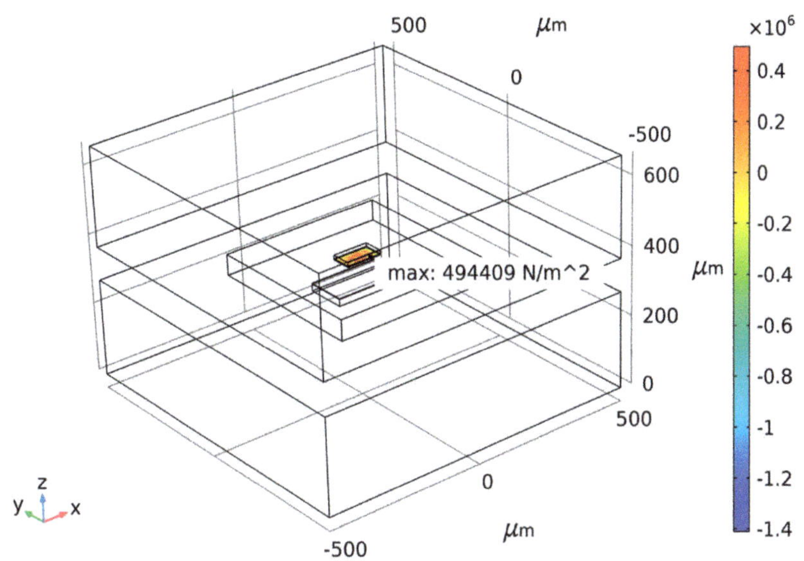

Figure 7. The maximum normal stress diagram of Au-Si anchor zone at 278.15 K with 0.5 anchor–zone ratio.

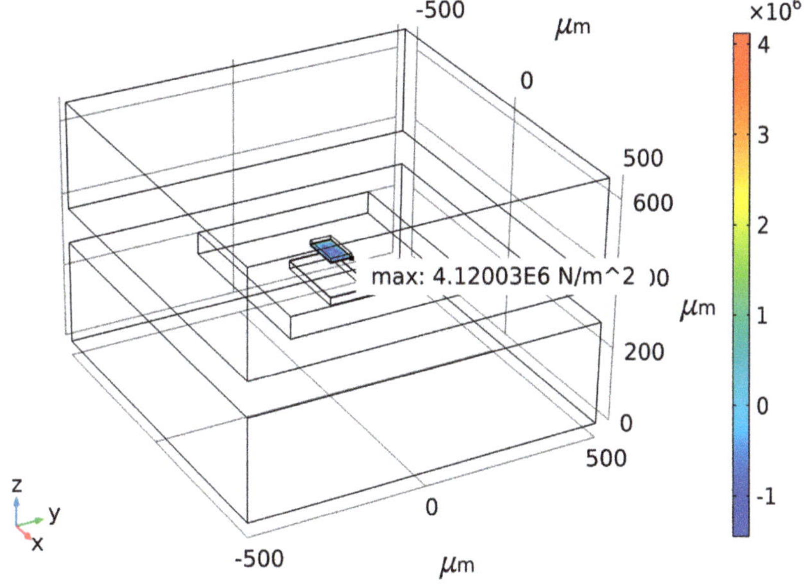

Figure 8. The maximum normal stress diagram of Au-Si anchor zone at 333.15 K with 0.5 anchor–zone ratio.

5. Design Verification

In order to verify the above theory, accelerometers with a 0.5 and 0.8 anchor–zone ratio were designed. The structure layout of the anchor area is shown in Figure 9. The anchor area of the Au-Si layer with an anchor–zone ratio of 0.5, as shown on the left, with an anchor area of 80 μm × 80 μm, compared with an anchor–zone ratio of 0.8 with an anchor area of 190 μm × 80 μm, as shown on the right.

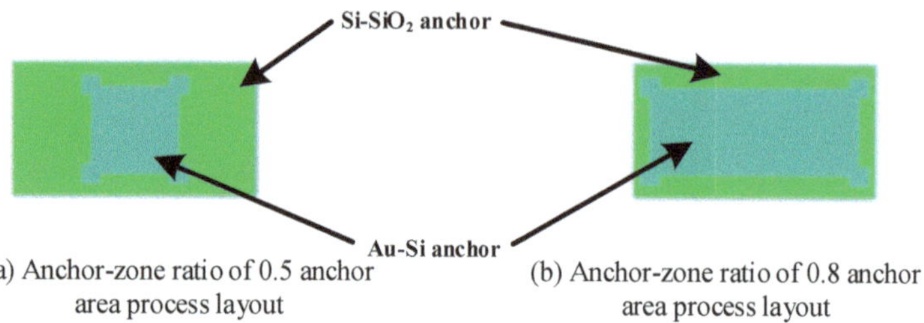

(a) Anchor-zone ratio of 0.5 anchor area process layout

(b) Anchor-zone ratio of 0.8 anchor area process layout

Figure 9. Anchor layout control group design drawing.

Both structures were processed using the standard unit, all-silicon wafer level packaging process, and their sensitive structures are identical. The two anchor structures obtained after processing are shown in Figure 10. The sensitive structures of the two accelerometers are exactly the same, except that the anchor–zone ratio is different. Figure 11 shows the SEM image of the sensitive structure layer of the MEMS accelerometer. Figure 12 is a picture of the chip after wafer level packaging, and Figure 13 is a picture of the final packaged MEMS accelerometer product.

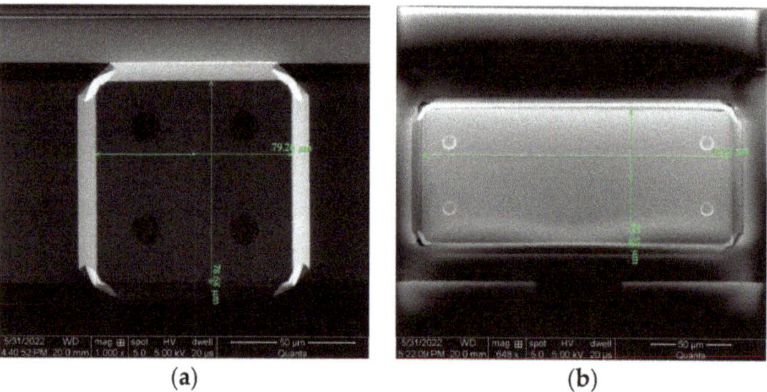

(a)

(b)

Figure 10. Two types of anchor SEM diagrams processed. (**a**) Anchor–zone ratio 0.5. (**b**) Anchor–zone ratio 0.8.

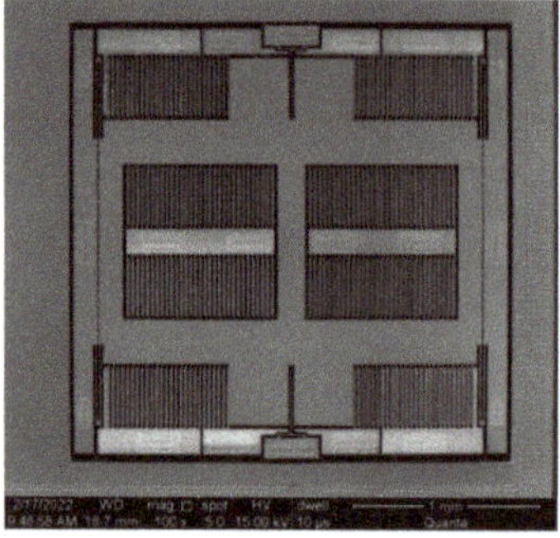

Figure 11. MEMS accelerometer SEM diagram of sensitive structural layers.

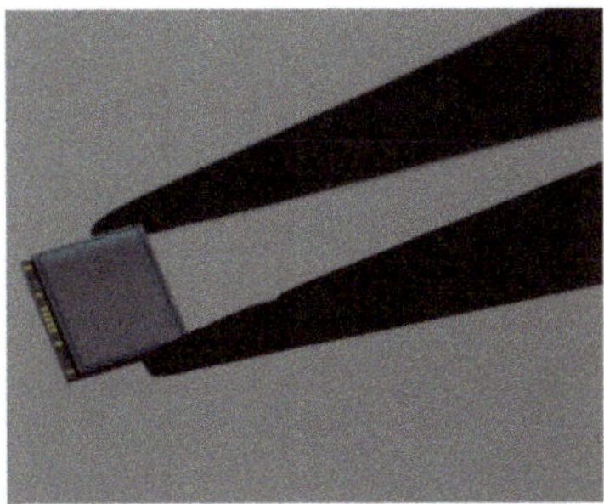

Figure 12. MEMS accelerometer structural chip photo.

Figure 13. Accelerometer product photo.

6. Experimental Test

The experiment involved testing the full-temperature performance of four MEMS accelerometers with anchor–zone ratios of 0.5 and 0.8. The accelerometer temperature performance test system is shown in Figure 14. Results from the two groups of eight accelerometers are presented in Figures 15–18. It can be observed that for the anchor–zone ratio of 0.8, the average zero-bias stability for acceleration is 133 μg, and the average coefficient stability for full-temperature scale is 87 ppm. On the other hand, for the anchor–zone ratio of 0.5, the average zero-deviation stability for acceleration is 46 μg, and the average full-temperature scale factor stability is 32 ppm. Compared with the two schemes, the zero-offset full-temperature performance of the scheme with an anchor–zone ratio of 0.5 is more outstanding, which has a 34.6% improvement in zero-bias stability and a 36.8% improvement in scale factor stability. The results of the comparison experiment demonstrate the effectiveness of the stress cancellation method in the anchor area of the MEMS accelerometer.

Figure 14. Accelerometer temperature performance test system.

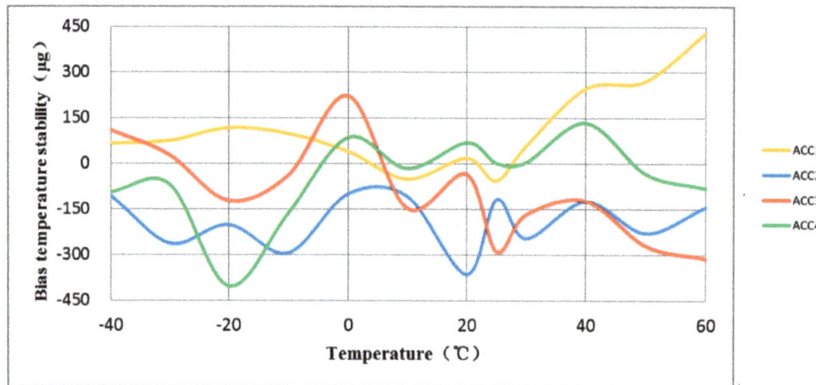

Figure 15. The average bias stability of the four accelerometers with an anchor-zone ratio of 0.8 was 133 µg.

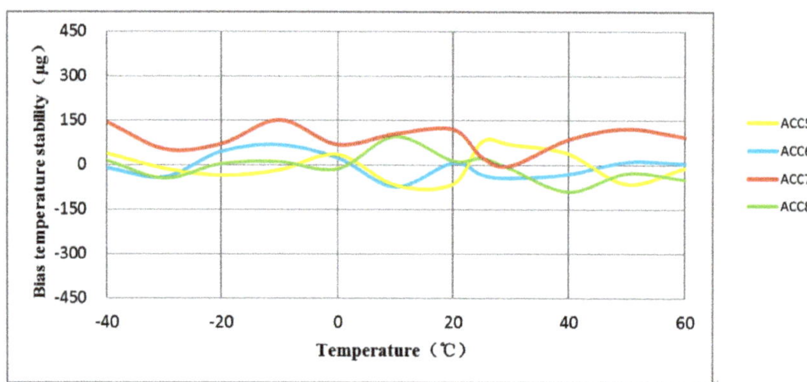

Figure 16. The average bias stability of the four accelerometers with an anchor-zone ratio of 0.5 was 46 µg.

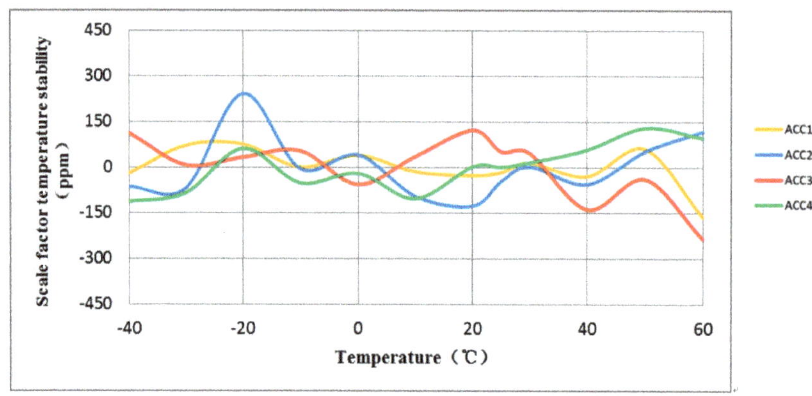

Figure 17. The average scale factor stability of the four accelerometers with an anchor-zone ratio of 0.8 was 87 ppm.

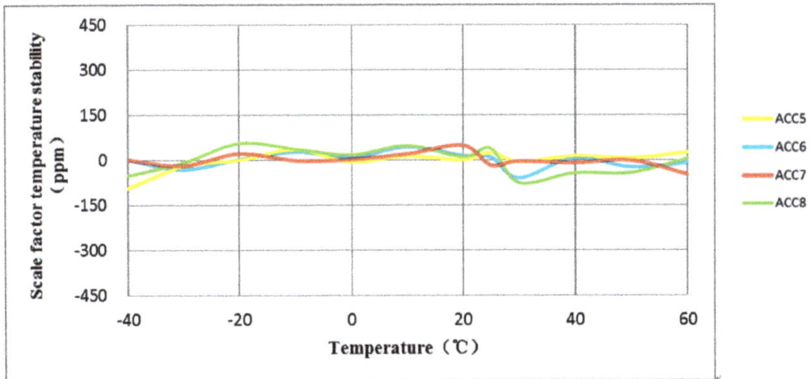

Figure 18. The average scale factor stability of the four accelerometers with an anchor-zone ratio of 0.5 was 32 ppm.

7. Conclusions

The simulation results illustrate that when the anchor–zone ratio of the accelerometer is reduced to 0.5, the stress in the anchor zone is significantly reduced, which improves the full-temperature performance of the accelerometer. Experiments show that if the anchor–zone ratio of the accelerometer is reduced from 0.8 to 0.5, the zero-bias full-temperature stability of the accelerometer is reduced from 133 μg to 46 μg, the scale factor full-temperature stability is reduced from 87 ppm to 32 ppm, the zero-bias full-temperature stability has a 34.6% improvement, and the scale factor full-temperature stability has a 36.8% improvement. The MEMS accelerometer has the characteristics of small size, light weight and low cost. The thermal expansion coefficient of each layer material is inconsistent. Thus, thermal stress will be generated due to the effects of temperature, which will affect the total temperature performance of the accelerometer. In order to reduce the influence of thermal stress on the multilayer structure, this paper presents a method of stress cancellation in the anchor zone, in order to reduce the internal thermal stress of structure. The experimental results show that when the anchor–zone ratio decreases from 0.8 to 0.5, the zero-offset stability of the MEMS accelerometer has been improved to 34.6%, and the full-temperature scale factor stability has been improved to 36.8%. The whole temperature performance of accelerometer is improved.

Author Contributions: Writing—original draft preparation, G.L. and Y.L.; writing—review and editing, Y.L., X.M. and X.Z.; technical supervision, X.W. and Z.J. All authors have read and agreed to the published version of the manuscript.

Funding: The work was supported by the National Natural Science Foundation of China (No. 51976177), common technology for equipment pre-research (50917020302), and Zhejiang Provincial Basic Public Welfare Research Program of China under Grant No. LGG20F040001.

Data Availability Statement: Not applicable.

Conflicts of Interest: The authors declare no conflict of interest.

References

1. Ru, X.; Gu, N.; Shang, H.; Zhang, H. MEMS Inertial Sensor Calibration Technology: Current Status and Future Trends. *Micromachines* **2022**, *13*, 879. [CrossRef] [PubMed]
2. Yazdi, N.; Najafi, K. An all-silicon single-wafer micro-g accelerometer with a combined surface and bulk micromachining process. *J. Microelectromech. Syst.* **2000**, *9*, 544–550. [CrossRef] [PubMed]
3. Marjoux, D.; Ullah, P.; Frantz-Rodriguez, N.; Morgado-Orsini, P.F.; Soursou, M.; Brisson, R.; Lenoir, Y.; Delhaye, F. Silicon MEMS by Safran—Navigation grade accelerometer ready for mass production. In Proceedings of the 2020 DGON Inertial Sensors and Systems (ISS), Braunschweig, Germany, 15–16 September 2020.

4.   Xiao, D.; Xia, D.; Li, Q.; Hou, Z.; Liu, G.; Wang, X.; Chen, Z.; Wu, X. A double differential torsional accelerometer with improved temperature robustness. *Sens. Actuators A Phys.* **2016**, *243*, 43–51. [CrossRef]

5.   Xu, W.; Tang, B.; Xie, G.; Yang, J. An All-Silicon Double Differential MEMS Accelerometer with Improved Thermal Stability. In Proceedings of the 2018 IEEE SENSORS, New Delhi, India, 28–31 October 2018.

6.   Niu, H.; Sun, G.; Wang, S.; Zhang, F. A design of capacitance MEMS accelerometer with wafer level encapsulated all-silicon comb tooth. *J. Chin. Inert. Technol.* **2020**, *28*, 672–676.

7.   Liu, H.; Fang, R.; Miao, M.; Zhang, Y.; Yan, Y.; Tang, X.; Lu, H.; Jin, Y. Design, Fabrication, and Performance Characterization of LTCC-Based Capacitive Accelerometers. *Micromachines* **2018**, *9*, 120. [CrossRef] [PubMed]

8.   Aydemir, A.; Akin, T. Self-Packaged Three Axis Capacitive Mems Accelerometer. In Proceedings of the 2020 IEEE 33rd International Conference on Micro Electro Mechanical Systems (MEMS), Vancouver, BC, Canada, 18–22 January 2020; pp. 777–780.

9.   He, Y.; Si, C.; Han, G.; Zhao, Y.; Ning, J.; Yang, F. A Novel Fabrication Method for a Capacitive MEMS Accelerometer Based on Glass–Silicon Composite Wafers. *Micromachines* **2021**, *12*, 102. [CrossRef] [PubMed]

10.   Zhou, Y.; Cao, H.; Guo, T. A Hybrid Algorithm for Noise Suppression of MEMS Accelerometer Based on the Improved VMD and TFPF. *Micromachines* **2022**, *13*, 891. [CrossRef] [PubMed]

11.   Yin, Y.; Fang, Z.; Dong, J.; Liu, Y.; Han, F. A Temperature-Insensitive Micromachined Resonant Accelerometer with Thermal Stress Isolation. In Proceedings of the 2018 IEEE SENSORS, New Delhi, India, 28–31 October 2018.

12.   Fujiyoshi, M.; Kawamoto, A.; Hashimoto, S.; Omura, Y.; Funabashi, H.; Ohira, Y.; Hata, Y.; Ozaki, T.; Akashi, T.; Nonomura, Y.; et al. Stress Isolation Suspension for Silicon-on-Insulator 3-Axis Accelerometer Designed by Topology Optimization Method. *IEEE Sens. J.* **2022**, *22*, 3965–3973. [CrossRef]

13.   Cui, J.; Liu, M.; Yang, H.; Li, D.; Zhao, Q. Temperature Robust Silicon Resonant Accelerometer with Stress Isolation Frame Mounted on Axis-Symmetrical Anchors. In Proceedings of the 2020 IEEE 33rd International Conference on Micro Electro Mechanical Systems (MEMS), Vancouver, BC, Canada, 18–22 January 2020.

14.   Schroder, S.; Niklaus, F.; Nafari, A.; Westby, E.R.; Fischer, A.C.; Stemme, G.; Haasl, S. Stress-minimized packaging of inertial sensors by double-sided bond wire attachment. *J. Microelectromech. Syst.* **2015**, *24*, 781–789. [CrossRef]

15.   Zhang, Y.; Gao, C.; Meng, F.; Hao, Y. A SOI sandwich differential capacitance accelerometer with low-stress package. In Proceedings of the 9th IEEE International Conference on Nano/Micro Engineered and Molecular Systems (NEMS), Waikiki Beach, HI, USA, 13–16 April 2014; pp. 341–345.

Article

High-Sensitivity Piezoelectric MEMS Accelerometer for Vector Hydrophones

Shuzheng Shi [1,2,†]**, Liyong Ma** [1,†]**, Kai Kang** [1]**, Jie Zhu** [3]**, Jinjiang Hu** [1,4]**, Hong Ma** [1]**, Yongjun Pang** [1,*] **and Zhanying Wang** [1,*]

[1] School of Mechanical Engineering, Hebei University of Architecture, Zhangjiakou 075000, China; shishuzheng2000@163.com (S.S.); maliyong@buaa.edu.cn (L.M.); kangkai_jgxy@163.com (K.K.); 15933136690@163.com (J.H.); mahong0210@163.com (H.M.)

[2] HBIS Group Co., Ltd., Shijiazhuang 050023, China

[3] School of Computer Science and Engineering, North China Institute of Aerospace Engineering, Langfang 065000, China; zhujie0424@126.com

[4] School of Materials Science and Engineering, University of Science and Technology Beijing, Beijing 100083, China

[*] Correspondence: pyj1063@hebiace.edu.cn (Y.P.); wzy0313@126.com (Z.W.)

[†] These authors contributed equally to this work.

Abstract: In response to the growing demand for high-sensitivity accelerometers in vector hydrophones, a piezoelectric MEMS accelerometer (PMA) was proposed, which has a four-cantilever beam integrated inertial mass unit structure, with the advantages of being lightweight and highly sensitive. A theoretical energy harvesting model was established for the piezoelectric cantilever beam, and the geometric dimensions and structure of the microdevice were optimized to meet the vibration pickup conditions. The sol-gel and annealing technology was employed to prepare high-quality PZT thin films on silicon substrate, and accelerometer microdevices were manufactured by using MEMS technology. Furthermore, the MEMS accelerometer was packaged for testing on a vibration measuring platform. Test results show that the PMA has a resonant frequency of 2300 Hz. In addition, there is a good linear relationship between the input acceleration and the output voltage, with $V = 8.412a - 0.212$. The PMA not only has high sensitivity, but also has outstanding anti-interference ability. The accelerometer structure was integrated into a vector hydrophone for testing in a calibration system. The results show that the piezoelectric vector hydrophone (PVH) has a sensitivity of -178.99 dB@1000 Hz (0 dB = 1 V/μPa) and a bandwidth of 20~1100 Hz. Meanwhile, it exhibits a good "8" shape directivity and consistency of each channel. These results demonstrate that the piezoelectric MEMS accelerometer has excellent capabilities suitable for use in vector hydrophones.

Keywords: microelectromechanical systems; cantilever beam; piezoelectric accelerometer; vector hydrophone; sensitivity

Citation: Shi, S.; Ma, L.; Kang, K.; Zhu, J.; Hu, J.; Ma, H.; Pang, Y.; Wang, Z. High-Sensitivity Piezoelectric MEMS Accelerometer for Vector Hydrophones. *Micromachines* **2023**, *14*, 1598. https://doi.org/10.3390/mi14081598

Academic Editor: Ha Duong Ngo

Received: 19 June 2023
Revised: 8 August 2023
Accepted: 11 August 2023
Published: 14 August 2023

1. Introduction

Accurately extracting information from underwater acoustic sources is a crucial aspect of sonar system research. Hydrophones serve as a window for communication between sonar systems and the ocean, acting as the "eyes and ears" of the sonar system. A hydrophone is a transducer to convert the acoustic radiation signals of underwater targets into electrical signals, which is commonly used for transmitting and receiving acoustic signals [1–3]. A vector hydrophone can measure vector information, such as the direction of underwater targets, particle velocity, displacement, and acceleration of water particles, while maintaining spatial colocation and time synchronization [4,5]. Due to its excellent sensitivity and directional capabilities, the vector hydrophone is widely used for detecting low-frequency underwater target signals. According to the measurement principle, the vector hydrophone can be divided into two types: pressure difference type [6] and resonance type [7]. Compared to the pressure difference type, the resonance type sensor can directly measure the movement of the sound wave and has advantages such as reliable performance, high sensitivity, and good low-frequency directivity. A better vibration design scheme is to install a sensor along the orthogonal coordinate axis inside a rigid shell to pick up the vibration signal of the water particle. The accelerometer is suitable for a vector hydrophone, because its output characteristic is dipole, and it is possible to measure low frequencies. It requires that the vibration pickup should have a small enough mass to approach neutral buoyancy and resonate with the water medium to pick up underwater vibration signals. In 2023, Mireles, J. et al. [8] developed a MOEMS accelerometer using SOI technology, which contains a mass structure and handle layers coupled with four designed springs built. The sensor exhibited a resonant frequency of 1274 Hz under running conditions up to 7 g. However, the sensitivity of the low-frequency response is lower, as the sensitivity is reduced by 6 dB for each doubling of the frequency. Therefore, a resonance type vector hydrophone requires a lightweight, compact, and highly sensitive micro accelerometer as the vibration pickup unit [9]. The MEMS accelerometer is smaller than a traditional accelerometer, reducing the impact of the original acoustic field when used in the manufacture of hydrophones, and making the detection more accurate. Meanwhile, it has lighter weight, lower manufacturing cost, and easier-to-achieve low-frequency target detection [10].

Currently, there are three types of MEMS accelerometers: piezoresistive [11], capacitive [12], and piezoelectric [13]. The piezoresistive MEMS accelerometer is an early developed type of miniaturized accelerometer. It consists of a cantilever beam with a piezoresistive material and an inertial mass unit, which possesses the advantages of simple structure and relatively easy fabrication. However, the large temperature coefficient of the piezoresistive material results in the sensitivity of the acceleration detection performance to temperature and the poor anti-interference ability. Additionally, it has low sensitivity, significant hysteresis effects, and a high-power dependence, making it difficult to apply to underwater scenarios with high requirements for sensitivity and accuracy [14]. The capacitive MEMS accelerometer uses variable capacitors as sensing components and has the advantages of high sensitivity, simple structure, low noise, and low temperature sensitivity, whereas it is susceptible to parasitic capacitance interference and requires complex detection circuits. Additionally, the non-linear effect of the capacitive accelerometer structure is significant at large displacement, limiting its linear dynamic testing range [15]. Unlike piezoresistive and capacitive sensors, the piezoelectric MEMS accelerometer is a passive device that requires no external power supply. It exhibits excellent piezoelectric coupling, high quality factor, wider linear amplitude range, and lower power consumption, making it widely used in marine resource exploration and inertial sensing systems [16]. Piezoelectric MEMS accelerometers commonly use lead zirconate titanate (PZT), zinc oxide (ZnO), and aluminum nitride (AlN) as thin film materials for electromechanical conversion. As three kinds of typical piezoelectric materials, the piezoelectric coefficients d_{31} of AlN and ZnO thin film is 3.4~6.4 pC/N and 5.9~12.4 pC/N, respectively [17]. However, the piezoelectric coefficients d_{31} of PZT thin film with high energy density and output voltage is up to 60~223 pC/N [18], tens of times higher than the value of AlN and ZnO. PZT has gained significant interest due to its high dielectric constant, low cost of fabrication, good piezoelectric properties, and convenience to be incorporated into MEMS accelerometers. Several studies have been carried out on piezoelectric MEMS accelerometers. Between 2006 and 2010, Hinichsen et al. [19–21] developed a series of PZT piezoelectric accelerometers based on beam structure and the inertial mass unit. Firstly, they fabricated and characterized a micromachined accelerometer with four beams and four inertial unit structures. Subsequently, they presented a theoretical model of a PZT MEMS accelerometer based on four-cantilever beams and single inertial unit. In 2019, Xu et al. [22] fabricated a PZT four-cantilever beams accelerometer with interdigital electrodes structure. The results indicated that a higher voltage output can be achieved with a smaller filling and electrode interspace. Recently, in 2021, Lee et al. [23] proposed and fabricated a piezoelectric MEMS accelerometer on Si substrates using photolithography and sol-gel PNZT. The accelerometer exhibited a sensitivity of 16.8 mV/g at 200 Hz. According to the above research, the four-cantilever beam structure has advantages, such as high sensitivity, wide frequency range, improved linearity, low power consumption, and easy fabrication, making it a popular choice for piezoelectric accelerometers used in various applications [8].

Research has focused on the theoretical analysis and functional verification of piezoelectric MEMS accelerometers in vector hydrophones to improve their design and performance. In 2017, Kim et al. [24] developed a PMN-PT crystal vibrator for vector hydrophones to measure the magnitude and direction of acoustic signals. The voltage response equation of the PMN-PT crystal vibrator was derived to guide the design of shear moding accelerometers in vector hydrophones. However, the accelerometer was not packaged as a hydrophone for testing and verification. In 2018, Seonghun et al. [25] proposed a shear-type accelerometer for a vector hydrophone, which was applied to a towed array sonar system. The minimum receiving voltage sensitivity (RVS) of the hydrophone demonstrated a remarkable capability of up to −201.4 dB at the operating frequency range, achieving the highest RVS over the frequency range. It should be noted that a single hydrophone can only test directivity in one direction. In 2020, Cho et al. [26] developed a PIN-PMN-PT piezoelectric accelerometer for vector hydrophone. Results showed the sensitivity of the accelerometer and hydrophone to be −199 dB and −196 dB, respectively. The sensor produced the expected cardioid directivity pattern across the operating frequency range. Although the basic functionality of the proposed vector sensor was confirmed, further research should be conducted to enhance its sensitivity. In 2022, Roh et al. [27] designed and optimized a shear-type accelerometer to improve the performance of vector hydrophones. The RVS of the hydrophone is −204.9 dB. Coupled with an omnidirectional hydrophone, the dipole response can generate a cardioid directivity pattern that detects both the magnitude and direction of external sound sources. Although significant studies have been made in the design and development of vector hydrophones, there are still challenges that need to be addressed. For instance, more accelerometer research is required to increase the sensitivity of vector hydrophones, particularly at a low-frequency range. In addition, the translation of theoretical designs into viable sensors is still a hurdle that needs to be overcome.

Therefore, based on a cross-beam integrated inertial mass unit structure, this paper developed a piezoelectric MEMS accelerometer (PMA) for vector hydrophones. Firstly, the impact of the geometric dimensions of the cantilever beam on resonance frequency and stress distribution was numerically analyzed, and its parameters were optimized. In addition, the overall structure of the piezoelectric microstructure was designed, and the manufacturing process of the device was proposed. PZT piezoelectric films were prepared using sol-gel technology as the sensitive unit, and MEMS technology was used to manufacture the acceleration microdevice. The MEMS accelerometer was packaged on a vibration test platform in order to investigate and analyze the performance of the PMA within a range of input acceleration and frequency. Finally, the accelerometer was encapsulated as a piezoelectric vector hydrophone (PVH) to investigate the sensitivity and directivity in a calibration system. The PVH with PMA has high sensitivity and good directivity.

2. Theoretical

Piezoelectric materials play an important role in the sensor dynamic response. Under different external forces, piezoelectric materials exhibit different types of electromechanical conversion modes, as shown in Figure 1. Most piezoelectric materials used in energy conversion sensors have a clear polar axis, and the different direction between external forces and the polar axis affects the material performance [28]. According to the working mode, the piezoelectric system mainly includes d_{31} working mode and d_{33} working mode. In the d_{31} mode, the external forces applied to the piezoelectric material are perpendicular to the electric field polarization direction. In other words, direction 1 is the direction of generating the coupled electric field, and direction 3 is the direction of applying the force, forming a sandwich structure with top and bottom electrodes. In the d_{33} mode, the piezoelectric material is subjected to force, and the polarized electric field is along the direction 3, forming a plane–bimorph structure. In the d_{31} and d_{33} working modes, the charge in the piezoelectric layer is induced to be perpendicular and parallel to the strain direction, respectively. The open-circuit voltage V_{oc} and transferred charge Q_{oc} of the piezoelectric layer can be expressed as follows [29,30]:

$$\begin{cases} V_{oc} = \frac{\sigma_{3i} d_{3i} t}{\varepsilon_r \varepsilon_0} \\ Q_{oc} = -\sigma_{3i} d_{3i} A_{3i} \end{cases} \tag{1}$$

where σ_{3i} ($i = 1, 3$) represents stress; t represents the piezoelectric layer thickness; d_{3i} ($i = 1$, 3) represents the piezoelectric coefficient; ε_r and ε_0 represent the relative dielectric constant and vacuum permittivity, respectively; and A_{3i} ($i = 1, 3$) represents the effective area of the electrodes.

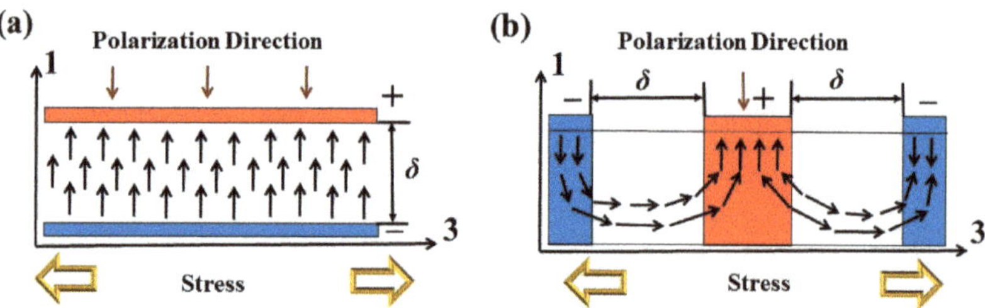

Figure 1. Electromechanical conversion mode of piezoelectric material: (**a**) longitudinal mode d_{31} and (**b**) transverse mode d_{33}.

In Equation (1), V_{oc} is proportional to σ_{3i}, d_{3i}, and t, and Q_{oc} is proportional to σ_{3i}. Obviously, the piezoelectric material performance depends on the type of working mode. Normally, the piezoelectric film is usually very thin to reduce the geometric dimensions of the sensor structure. The thickness of the piezoelectric film in d_{31} is shorter than d_{33}. The d_{31} working mode has the advantage of higher voltage output, while d_{33} has a larger current output. Meanwhile, the d_{31} PZT film will produce larger mechanical strain during vibration.

Due to the strain generated inside the cantilever beam, which is a relatively simple structure during vibration, the piezoelectric cantilever beam is the key structure for collecting mechanical energy from vibration. According to the piezoelectric working mode and mechanical vibration, an external force can drive the cantilever beam to vibrate. Vibration causes polarization changes in the piezoelectric film on the cantilever beam, resulting in voltage signals on the surface of the thin film. The compressive stress on the upper surface and the tensile stress on the lower surface generate an electric charge of opposite polarity on both surfaces of the film, causing opposite deformation of the two parts of a single cantilever beam. The stress difference on the two surfaces of the cantilever beam leads to the distribution of charges between the two surfaces, as shown in Figure 2. In this structure, there are two piezoelectric units connected in series, one lower electrode and two disconnected upper electrodes [31].

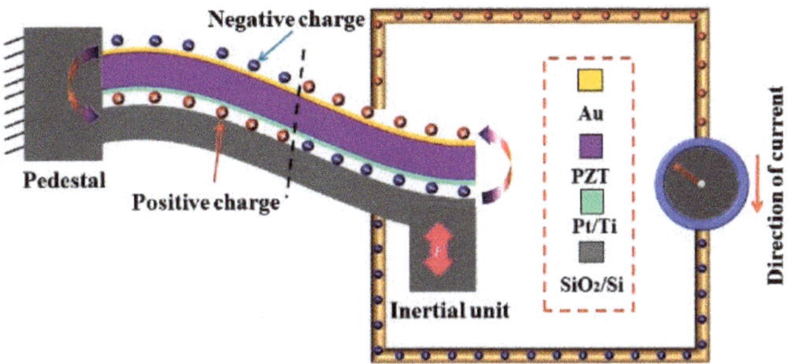

Figure 2. The electrical charges generated on the single cantilever beam.

3. Design and Simulation

3.1. Design of the Sensor Microstructure

According to the above theory, we designed a MEMS accelerometer with four cantilever beams and a central inertial unit based on Au/PZT/Pt/Si, as shown in Figure 3a. The vibration energy is converted into electricity by the d_{31} PZT film. When the central inertial unit is subjected to vibration, the cantilever beam experiences stress in the opposite direction, generating an electrical signal with opposite phase. This multichannel output improves the stability of the signal from the device. The upper and lower electrodes are made of low-resistivity Au and Pt, respectively. Due to the constraints of the four-beam structure, when the transverse swing of the mass unit is ignored, acceleration triggers the vibration of the inertial mass unit. This vibration drives the cantilever beam to swing, resulting in the generation of strain in the PZT film integrated on the suspended beam, which then produces an electrical signal output reflecting the value of the acceleration. Similar to other forms of accelerators, mechanical properties are determined by the structural parameters of the beam. To establish a Cartesian coordinate system (o-x-y-z), shown in Figure 3b, the length, width, and thickness of the beam are denoted by l, w, and t, respectively. The central coordinate of the inertial unit is located at $x = l$. In the theoretical model, the bending of the beam mass unit and inertial mass unit is ignored, and the effect of mass unit motion without rotation is guaranteed. Without considering the gravitational acceleration, the mass unit m will move downward or upward under an external acceleration a_z in the z-axis direction. The four cantilever beams undergo the same deformation due to the symmetry of the microstructure. In accordance with small deflection theory, the torque balance equation on the beam is expressed as [32]:

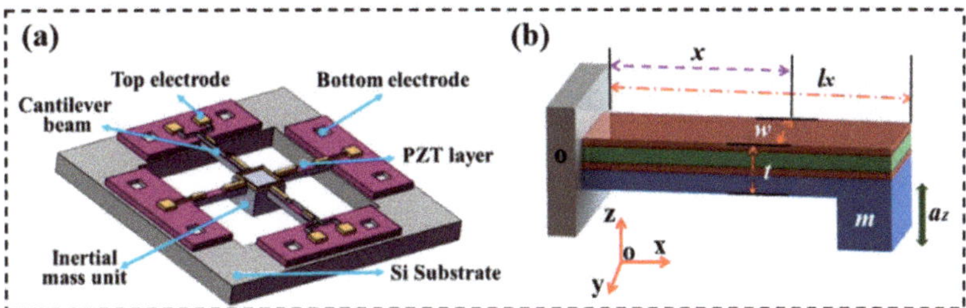

Figure 3. Microstructure model: (**a**) the microstructure with four cantilever beams and an inertial mass unit, (**b**) coordinates and parameters of a single cantilever beam.

$$EI_x\omega''(x) = F(l_x - x) - M_0 \tag{2}$$

where x represents the distance from the constraint end to any point on the beam. E is Young's modulus of Si, I_x is the moment of inertia along the x-axis, $\omega(x)$ is the displacement of l_x along the z-axis, F is the reaction force, and M_0 is restriction moment.

$$I_x = wt^3/12 \tag{3}$$

The concentrated force F on the cantilever beam can be determined as:

$$F = ma_Z/4 \tag{4}$$

where m represents the mass of the inertia unit.

$$\omega(x) = \int \int_0^{l_x} \left(\frac{3ma_Z}{EI} dx \right) dx + Ax + B \tag{5}$$

A and B are constants. The boundary condition is expressed as:

$$\omega(0) = \omega'(0) = 0 \tag{6}$$

From Equations (3), (5), and (6), the displacement of l_x along the z-axis direction is expressed as:

$$\omega(x) = \frac{12ma_Z}{4Ewt^3} \left(\frac{1}{2}l_x x^2 - \frac{1}{6}x^3 \right) \tag{7}$$

When $x = l_x$, the displacement can be expressed as:

$$\omega(l_x) = \frac{ma_Z l_x{}^3}{Ewt^3} \tag{8}$$

In the x-axis direction, the maximum stress of the cantilever beam can be expressed as:

$$\sigma_{\max} = \frac{Fd}{I_x} = \frac{3l_x}{2wt^2}ma_Z \tag{9}$$

where $d = h/2$ is the distance between the neutral plane of the cantilever beam and the piezoelectric layer. According to Hooke's law, the displacement of the inertial unit is equal to the displacement of the cantilever. The elastic constant can be expressed as:

$$\left(\frac{k}{m} \right)^2 = \frac{ma_Z}{\omega(x)} = \int_0^{l_x} EI[\omega''(x)]^2 dx / \frac{1}{2} m[\omega(l_x)]^2 \tag{10}$$

$$\frac{k}{m} = \sqrt{2Ewt^3/4ml_x^3} \tag{11}$$

where k represents the elastic coefficient of the cantilever beam. The resonance frequency of the cantilever beam can be expressed as:

$$f = \frac{1}{2\pi}\sqrt{\frac{k}{m}} = \frac{1}{2\pi}\sqrt{Ewh^3/4ml_x^3} \tag{12}$$

Positive and negative charges appear on the top and bottom surfaces of the PZT film, and the charge quantity q is expressed as:

$$q = d_{31}\sigma_{\max} \tag{13}$$

where the d_{31} is the piezoelectric coefficient of normal stress, $\sigma_{\max}$ is the normal stress. In light of the definition, the capacitance C between the top and bottom electrode can be obtained as follows:

$$C = q/V \tag{14}$$

By substituting Equations (9) and (13) into Equation (14), the output voltage V is expressed as:

$$V = d_{31} \frac{3l_x}{2wt^2C} ma_z \tag{15}$$

Combined with Equations (14) and (15), the sensitivity S of the PZT element can be expressed as:

$$S = \frac{\Delta V}{\Delta a} = d_{31} \frac{3l_x m}{2wt^2C} \tag{16}$$

According to the above analysis, to improve the output voltage sensitivity, it is recommended to increase the inertial unit mass and beam length, while decreasing the beam thickness and width. During the size determination of the MEMS accelerometer, factors include the microdevice sensitivity and microstructure working frequency bandwidth. Microdevice sensitivity is proportional to the cantilever beam stress, with higher stress leading to greater sensitivity and vice versa. Broaden the microstructure working frequency bandwidth by increasing the natural frequency of the sensitive microdevice, but fixed structure size means the sensitivity and frequency are contradictory. PVH performance bandwidth is mainly determined by the natural frequency of the sensing structure, and increased natural frequency leads to the wider bandwidth of PVH. Hence, it is necessary to conduct further static stress to determine the sensor dimensions.

3.2. Simulation

To consider the performance of the microstructure before fabrication, Finite Element Analysis was performed using COMSOL Multiphysics 6.1 software, as shown in Figure 4. In the design of beam and inertial mass unit structure, not only the material settings, but also the physical settings of the Multiphysics simulation interface of the piezoelectric components are used; the simulation parameters of the principal material are listed in Table 1.

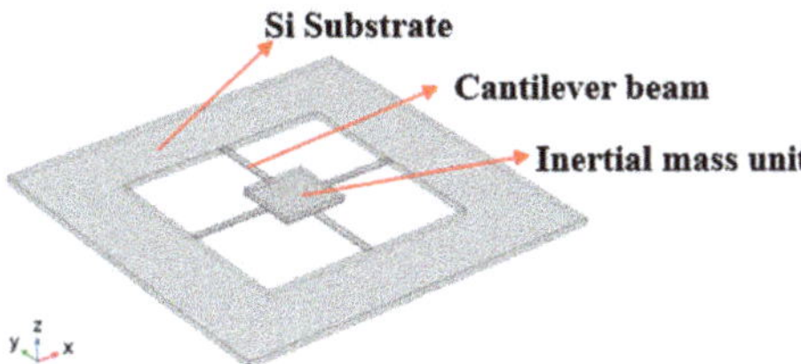

Figure 4. Microstructural model in COMSOL Multiphysics.

Design variables are the length of cantilever beams, widths of cantilever beams, thicknesses of cantilever beams, and the side length of the inertial mass unit. Under the action of the external force F_z, the relationship of the microstructure dimensions, the resonant frequency, and the stress is shown in Figure 5. Figure 5a shows the force analysis and deflection diagram of the microstructure. By varying the microstructure dimensions, which include the length, width, thickness of the cantilever beams, and the side length of the inertial mass unit, the resonant frequency and the maximum stress can be changed. Moreover, the effect of microstructure dimensions is being analyzed using COMSOL Multiphysics 6.1 through parametric scanning methods. Considering the restrictions of the manufacturing process and the sensor microstructure parameters, Table 2 shows the parameter setting for the scanning analysis. The other components are set as free moving parts, while the base is set as a fixed constraint. Figure 5b,c show that the increase of the width and thickness of the cantilever beam will increase the resonant frequency. On the contrary, the increase of the length of the cantilever beam or the inertial mass unit will keep the resonant frequency decreasing continuously. Using a parametric scanning simulation with the above parameters, the resonant frequency of approximately 4 kHz can be obtained. For an input acceleration of 1 g and a frequency of 1 kHz, the effects of the dimensions on the maximum stresses are shown in Figure 5d,e. The maximum stress decreases as the width and thickness of the cantilever beam increase, and increases as the inertial mass unit or side length of the beam increases. Although the maximum stress fluctuates, the overall trends of performance changes are consistent and well below the fatigue strength of the silicon substrate. From the analysis of the natural frequency and maximum stress, the microstructure dimensions can be determined as shown in Table 3.

Table 1. Simulation parameters of principal material.

Material	Parameters	Value
Si	Density	$2330 \, \text{kg·m}^{-3}$
	Young's modulus	190 Gpa
	Poisson's ratio	0.26
	Dielectric constant	11.9
PZT	Density	$7500 \, \text{kg·m}^{-3}$
	Young's modulus	75 Gpa
	Poisson's ratio	0.32
	Effective Coupling Co-efficient	$35 \, k^2\%$
	Curie temperature	$365 \,^\circ\text{C}$
	Elastic modulus	$5 \times 10^{10} \, \text{N/m}^2$
	Tensile modulus	$2 \times 10^{7} \, \text{N/m}^2$
	Dielectric constant	1300
	Piezoelectric charge constant d_{31}	$-270 \, \text{pC/N}$
	Mechanical quality factor	32

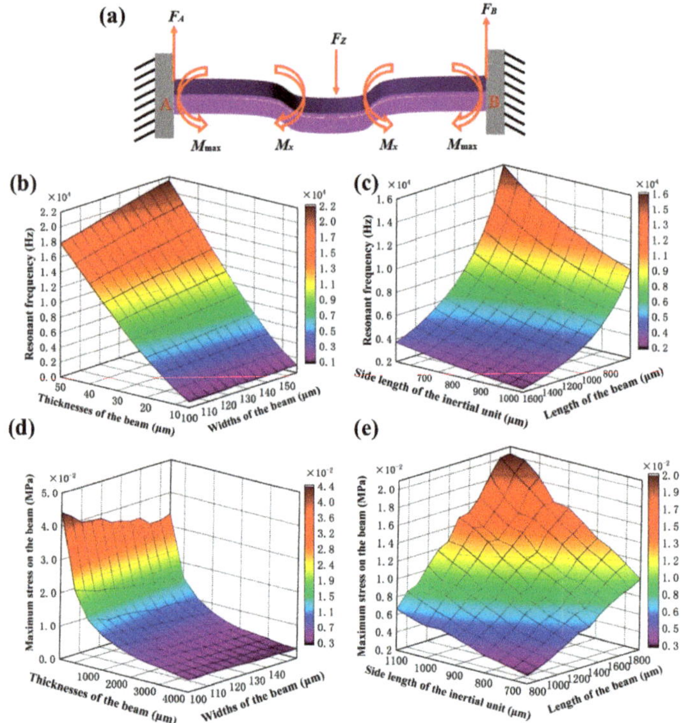

Figure 5. The effects of microstructural parameters on the resonant frequency and maximum stress: (**a**) the force analysis and deflection diagram of the microstructure, (**b**) relationship of the resonant frequency on the width and thickness of the beam, (**c**) relationship of the resonant frequency on the beam length and the inertial mass unit, (**d**) relationship of the maximum stress on the width and thickness of the beam, (**e**) relationship of the maximum stress on the beam length and the inertial mass unit.

Table 2. Parameters of the scanning.

Parameters	Range	Steps
Length of cantilever beams	700~1800	100 μm
Width of cantilever beams	100~150	5 μm
Thicknesses of cantilever beams	10~50	5 μm
Side length of inertial mass unit	600~1100	50 μm

Table 3. Dimension parameters of microstructure (Unit: μm).

Parameters for Scanning	Numerical Value
Length of single beams	1100 μm
Width of beams	120 μm
Thicknesses of beams	25 μm
Side length of the inertial unit	700 μm
Thicknesses of PZT	1 μm
Microstructure size	5000 μm × 5000 μm

When the geometric dimensions are obtained, the stress distribution and modal analyses can be performed in Figure 6. The acceleration of 1 g was loaded along the z-axis direction to the inertial mass unit; the stress distribution and deformation of the microstructure are showed in Figure 6a. It can be seen that the region of stress concentration is mainly distributed at both ends of the cantilever beam. Moreover, the stress near the support frame areas is more concentrated, which is in line with the expected design. Therefore, the piezoelectric unit should be arranged in the stress concentration area as much as possible to produce a larger output voltage. Figure 6b–d show that the 1st-, 2nd-, and 3rd-order resonance frequencies are 2298 Hz, 12,198 Hz, and 12,199 Hz, respectively. Whereas the 1st resonance mode exhibits vertical displacement, the 2nd and 3rd resonance modes exhibit torsional displacement in different directions. The working frequency of the accelerometer should be lower than the 1st-order resonance frequency to ensure stability of performance.

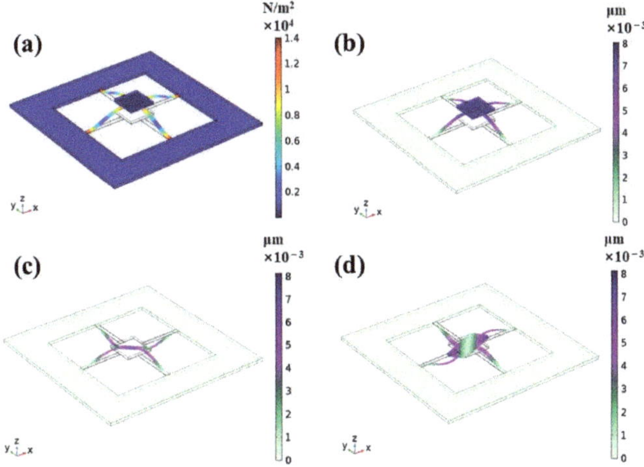

Figure 6. Modal analysis using COMSOL Multiphysics: (**a**) the deformation and stress distribution of the microstructure, (**b**) the 1st-ordered resonance, (**c**) the 2nd-ordered resonance, and (**d**) the 3rd-ordered resonance.

4. Fabrication and Testing

4.1. Fabrication of the MEMS Acceleration Microstructure

The acceleration microstructure was fabricated in a clean laboratory. The main processes include thermal oxidation, sputtering, sol-gel, photolithography, ion beam etching (IBE), and reactive ion etching (RIE). The fabrication scheme of the MEMS piezoelectric microstructure is shown in Figure 7.

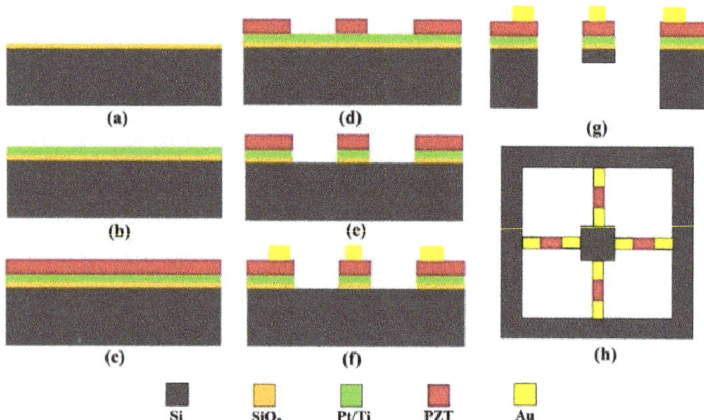

Figure 7. Fabrication scheme of the MEMS piezoelectric microstructure: (**a**) SiO$_2$ grown by thermal oxidation. (**b**) Pt/Ti sputtered onto the SiO$_2$/Si substrate. (**c**) PZT grown by sol-gel combined with the annealing process. (**d,e**) PZT, Pt/Ti, and SiO$_2$ pattern produced by photolithography and IBE. (**f**) Au/Ti was sputtered, followed by a peel-off process. (**g,h**) Cantilever beams and mass fabrication using lithography and etching by the RIE process.

I. Thermal oxidation was used to grow 330 nm SiO$_2$ on the silicon wafer surface as a transition layer (Figure 7a).

II. Pt/Ti was sputtered onto the SiO$_2$/Si substrate as the bottom electrode (Figure 7b).

III. A 1 µm thick PZT (PbZr$_{0.53}$Ti$_{0.47}$O$_3$) piezoelectric film was grown on the Pt/Ti/SiO$_2$/Si (100) substrate using sol-gel combined with the annealing process as the functional layer. The annealing process is completed in a tubular annealing furnace in an ultra-clean room. The annealing process is divided into three sections: the first section (low-temperature section) is 450 °C for 10 min; the second section (high-temperature section) is 650 °C for 10 min; and in the third section, 10% excess lead (Pb) is added at 700 °C for 30 min (supplement Pb lost in the annealing process). After annealing, a layer of PZT film with preferential crystallization growth is obtained, and more than 10 steps are conducted continuously to form 10 layers of PZT film (Figure 7c).

IV. The IBE process was used to sequentially etch PZT, Pt/Ti, and SiO$_2$ to pattern the PZT piezoelectric unit, Pt/Ti bottom electrode, and SiO$_2$ layer (Figure 7d,e).

V. Au/Ti was sputtered, followed by a peel-off process to complete the preparation of the top metal electrode on the PZT surface (Figure 7f).

VI. The RIE process was used to etch the front and back of the silicon to determine the thickness of the cantilever and release the four cantilevers (Figure 7g).

The acceleration microstructure manufactured through the above process is shown in Figure 7h. In the fabrication process of the microstructure, mature UV photolithography and dry etching techniques were used to ensure the accuracy of the structural dimensions of the cantilevers and the inertial mass unit as much as possible, which improved the stability of the sensor fabrication process and the yield of the device, laying a device foundation for the integration of hydrophones and the improvement of the sensor performance.

After the fabrication of PMA, scanning electron microscopy (FE-SEM, SUPPA-55, Germany, SE2, 14.8 nm, 10 kV, 26.79 KX) was used to obtain the accelerometer morphology in an ultra-clean room and room temperature, as shown in Figure 8. The microstructure has a clear and smooth boundary, and the surface is clean without stains. The cantilever beam has a width of 110.82 μm, which is consistent with the design dimensions. In addition, the small modulus of the cantilever beam is beneficial for the cantilever beam to bend under external vibration or impact force, thereby generating charges on different surfaces. The MEMS sensors ensure their long-term stability due to their sturdy mechanical structure. The microstructure is encapsulated on the tube shell at room temperature, and then pasted onto a printed circuit board (PCB) to form a device/tube shell/PCB packaging structure. The upper and lower electrodes are wire bonded separately to complete the packaging of the sensor. As shown in Figure 8a, the boundaries of the four cantilever beams are clear, and the structure is symmetrical. The metal electrode pattern is intact, and the inertial mass unit is in a suspended state, effectively avoiding interference between the mass unit and the bottom surface of the sensor during vibration. As can be seen from Figure 8b, the PZT film deposited on the cantilever beam has a thickness of about 1 μm, and the interface between each layer of the material is clearly defined. The PZT film is well bonded and adhered to the Pt/Ti/SiO$_2$/Si substrate, resulting in a good and clear interface between the PZT film and the substrate. In addition, the cantilever beam has a small modulus, which helps the beam to bend under external vibrations or impacts, thereby generating charges on different surfaces. PMA ensures long-term stability due to its sturdy mechanical structure.

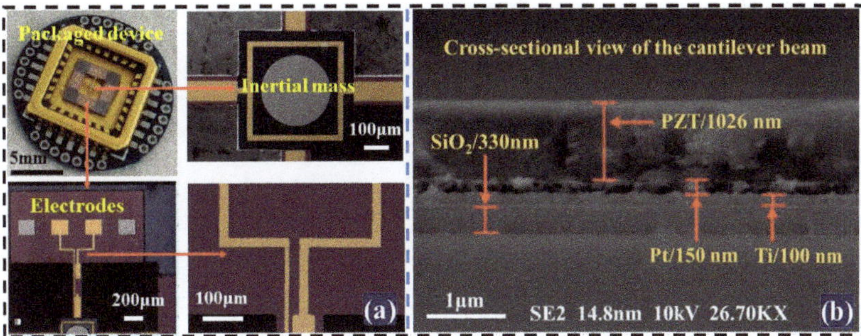

Figure 8. Morphological characterization of the PMA: (**a**) photos of the packaging equipment, indicating the electrodes on the base and cantilever beam, respectively, and (**b**) SEM images of cantilever beam cross-section, representing different thicknesses of PZT, Pt/Ti, SiO$_2$, and silicon wafer layers, respectively.

4.2. Performance of the MEMS Accelerometer

The vibration measurement system of the PMA is shown in Figure 9. The measuring platform is shown in Figure 9a. The vibration calibration system developed by B&K company is used to provide vibration interference of different frequencies for PMA. A function generator (Agilent 33522 A, Agilent Technologies, Santa Clara, CA, USA) generates a sine wave signal, which is amplified by a power 136 amplifier, and finally fed into the vibration exciter to generate the vibrations in the required directions [33]. The flow chart of measurement system is shown in Figure 9b. Both the packaged PMA and the standard accelerometer are fixed on the vibration platform to ensure that the maximum output direction is perpendicular to the platform surface. The test system generates a sine wave signal through the signal generator, which is enhanced by the power amplifier. The amplified signal drives the shaker to vibrate regularly, and the vibration frequency is consistent with the sine signal. After passing through the charge amplifier, the output voltage of the PMA is input to an oscilloscope to display the output voltage. Additionally, the power amplifier is set so that the vibration platform produces a certain amount of acceleration, as measured by the standard accelerometer. The fixed acceleration can be applied to the PMA to observe its output signal.

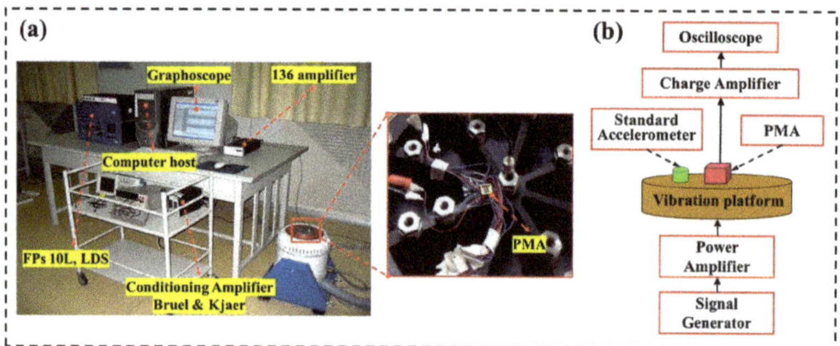

Figure 9. The vibration measurement system of the PMA: (**a**) vibration measuring platform, (**b**) flow chart of measurement system.

The resonance frequency of the PMA is shown in Figure 10. The output of the PMA increases with increasing frequency and reaches its maximum value at about 2300 Hz, indicating that this frequency is the resonant point of the PMA. After passing through the resonant point, the output voltage drops sharply. Nevertheless, there is still a certain difference between the measured and designed target frequency of about 3000 Hz. This difference may be attributed to the MEMS fabrication process, where there is a certain difference between the actual geometric dimensions and the designed ones of the cantilever beams and the central inertial unit. It is impossible to achieve a rigid installation between the sensor and vibration platform. In addition, the damping coefficient around the encapsulated acceleration sensor in air may also cause the frequency difference mentioned above [34]. In any case, the resonant frequency results prove the effectiveness of our designed accelerometer.

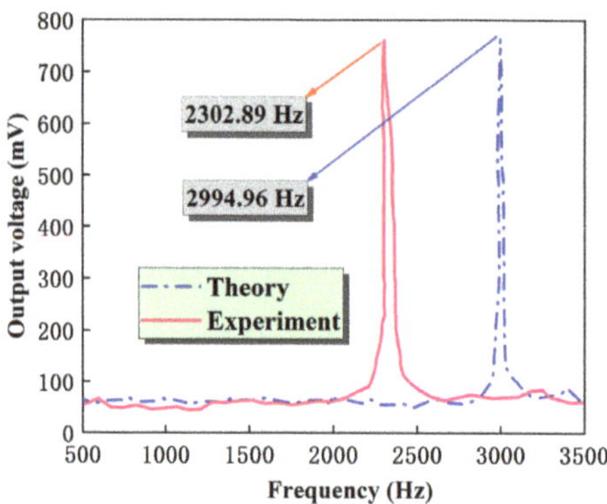

Figure 10. Resonant frequency curve of PMA.

A single cantilever beam was analyzed to study the noise impact of PMA that was measured on a vibration testing system. The accelerometer was installed on the excitation platform to test the sensor frequency response under load. On the other hand, the sensor was tested without being installed on the excitation table to measure its frequency response under no load. Under a 1000 Hz vibration frequency input, the results of the noise signal and output electrical signal comparison test are shown in Figure 11. The blue curve is for the no-load condition, and the peak-to-peak voltage value of the testing system is about 86 μV. The red curve is for the load condition, and the peak-to-peak voltage value of the piezoelectric element is about 183.4 mV, which is much larger than the output voltage value under no load, differing by about three orders of magnitude. The PMA is a passive piezoelectric sensor, and noise mainly comes from the testing system. The noise is relatively small compared to the output signal, so it can be ignored. The noise has little impact on the output signal of the piezoelectric element, demonstrating good anti-interference ability and reducing the influence of the noise gradient in the environment on the sensor detection performance.

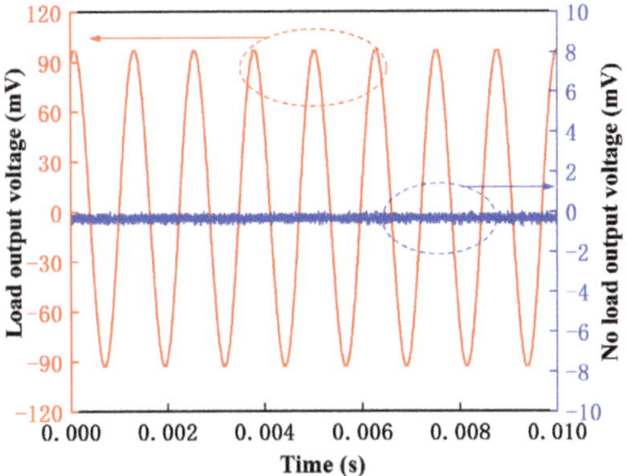

Figure 11. No-load and load output voltage curve.

In addition, the influence of different accelerations and frequencies on the acceleration performance is discussed. The voltage curve of the PMA is shown in Figure 12. By adjusting the power amplitude, the vibration platform was increased from 1 g to 5 g in steps of 1 g, and excited by a sine wave signal with input frequency at 100 Hz and 1000 Hz. Figure 12a,b are shown as output voltage curves of the PMA at 100 Hz and 1000 Hz, respectively. The results show that as the acceleration increases, the output signal is increased proportionally. The perfect response indicates the high-performance sensitivity of the PMA, which is able to meet the demand of hydrophones.

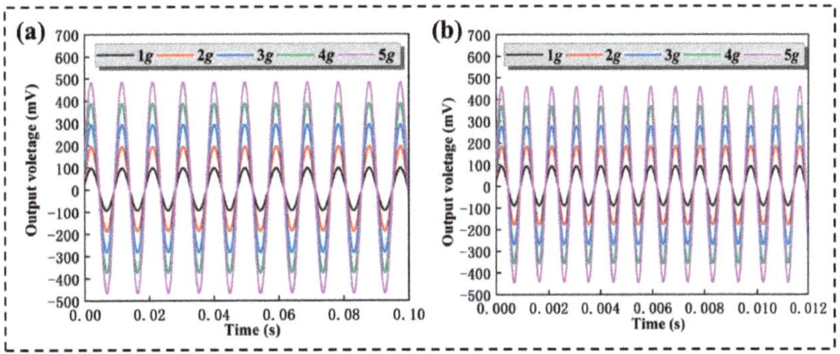

Figure 12. Output voltage curves of the PMA under different accelerations from 1 g to 5 g at (**a**) 100 Hz and (**b**) 1000 Hz.

Finally, sensitivity testing was performed on the accelerometer. The vibration platform was set to start from 0 g and gradually increase to 5 g in steps. Each acceleration value was loaded with a negative acceleration measurement for 2 min in the same way, i.e., applying −5~5 g acceleration to the sensor under testing. The output voltage results at each acceleration were exported from the data interface of the digital oscilloscope. Then, the data were linearly fitted using the least squares method, and Figure 13 shows the output curve of the test, which shows the relationship between the output voltage and PMA values. The relationship between the output voltage of a single cantilever beam and the acceleration value is plotted at an excitation frequency of 600 Hz. The results showed that the input acceleration and output voltage of the PMA had a linear relationship, and the function expression was $V = ka + V_o = 8.412a - 0.212$ (where k is the slope of the line, and V_o is the voltage intercept). The linear correlation coefficient R^2 of the sensor's single cantilever beam was 0.9997, approaching 1. Therefore, the sensor has good acceleration response linearity and passive stability. The slope is equal to its prime sensitivity, which measures the output voltage per unit of gravitational force applied. In this case, the slope is found to be 8.412 mV/g for a single beam. Hence, since all four beams on the accelerometer are coupled in parallel, the sensitivity of the entire sensor is 33.65 mV/g.

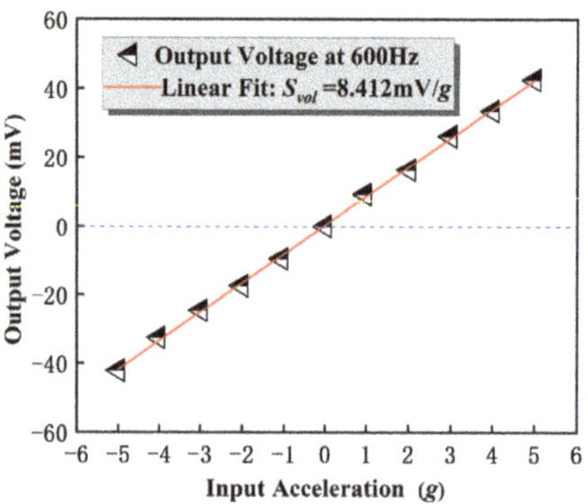

Figure 13. Output voltage sensitivity of the PMA at 600 Hz, ranging from −5 g to 5 g.

4.3. Performance of the MEMS Hydrophone

To investigate the performance of the PVH for vector hydrophone, which was encapsulated inside the polyurethane waterproof cap of the vector hydrophone [35], the standard wave tube calibration system was used to measure the sensitivity and directionality of the vector hydrophone [36], as shown in Figure 14. The standing wave tube measurement site is shown in Figure 14a. The PVH and a standard hydrophone were placed in a free sound field simultaneously. The output signals of the two sensors were compared with the standard hydrophone used as a reference to obtain the received sensitivity of the PVH. The sensitivity of the PVH was calibrated to the standard hydrophone (where sensitivity of the standard hydrophone is −180 dB (0 dB = 1 V/μPa)). The schematic diagram of the testing system is shown in Figure 14b. The PVH was installed on a mechanical rotating device, which had four orthogonal fixed ends for securing it. A sine wave with a peak-to-peak voltage of 1 V and 20 dB amplification was generated as the signal source, and the frequency range was swept from 1/3 octave to 2 kHz, with a sampling frequency of 20~2000 Hz. The elevation control console was adjusted to place the standard hydrophone and the uncalibrated hydrophone at distances d_0 and d from the water surface, respectively. The open-circuit out voltage e_0 of the standard hydrophone was measured to obtain its sound pressure p_0. Similarly, the open-circuit output voltage e_x of the uncalibrated hydrophone was measured, and the sensitivity of PVH at each frequency point was calculated based on the sound pressure information obtained at the location of the standard hydrophone and the two sets of measured values. According to the above analysis of the testing process, the values need to be recorded in the sensitivity calibration include: the open-circuit output voltage of PVH (e_x), the open-circuit output voltage of the standard hydrophone (e_0), and the calibration value of the standard hydrophone sensitivity in the free field (M_0). According to the definition of the hydrophone voltage sensitivity in the free field, M_0 is expressed as [4]:

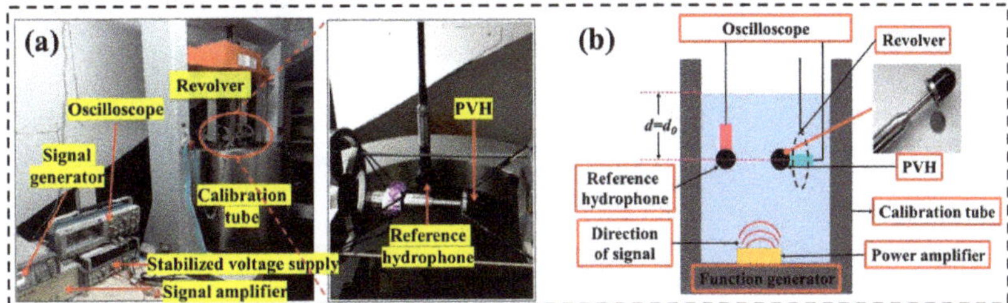

Figure 14. Measuring calibration system of the PVH: (**a**) test calibration system, (**b**) schematic diagram of hydrophone calibration device.

$$M_0 = e_0 / p_0 \tag{17}$$

The sensitivity of PVH is expressed as:

$$M_x = e_x / p_x \tag{18}$$

where p_0 and p_x are the sound pressures at the acoustic centers of the standard hydrophone and PVH, respectively, before placing them in the sound field. d and d_0 are the depths of the centers of the PVH and the standard hydrophone mounted on the mechanical rotation device in the water, respectively. The sound pressure at any point in the standing wave barrel satisfies the relationship $p \propto \sin kd$. In the testing process of this paper, the PVH and standard hydrophone are placed on the same horizontal plane ($d = d_0$). Therefore, the sensitivity of the PVH is given by:

$$M_x = 20\lg\left(\frac{e_x}{e_0} \frac{\sin kd}{\cos kd_0}\right) + M_0 \tag{19}$$

According to Equation (19), the sensitivity of the PVH can be obtained, and the sensitivity curve can be plotted based on the data. Figure 15a shows the x-axis and y-axis frequency response curve of the PVH. It can be seen that the PVH has good low-frequency characteristics, with a sensitivity of −178.99 dB@1000 Hz (0 dB = 1 V/μPa) and a frequency bandwidth of 20~1100 Hz, 11 dB higher than the bionic cilia MEMS vector hydrophone (CVH) at the same frequency [7], suitable for quiet underwater applications.

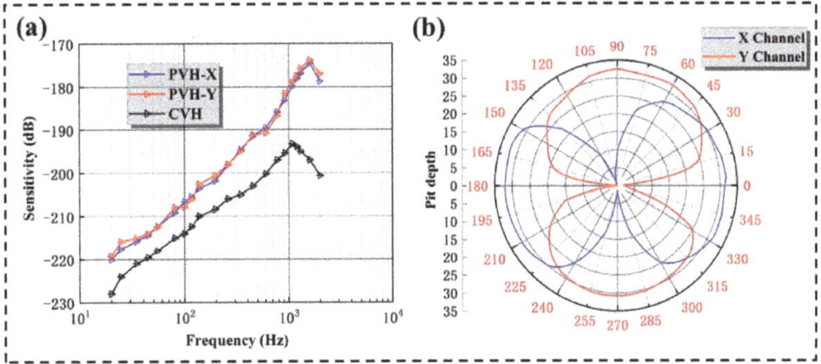

Figure 15. Measuring calibration system and performance of the PVH: (**a**) comparison chart of hydrophone sensitivity curve and (**b**) the "8" character directivity diagram at 800 Hz.

Theoretically, the various channels of a vector hydrophone should have cosine directivity independent of frequency. The PVH is suspended in a rotating frame within a standing wave tube, with the coordinate axis of the channel to be tested parallel to the axis of the standing wave tube and pointing towards the transmitting transducer (as is shown in Figure 14a). While keeping the output power and frequency of the transmitting transducer stable, the PVH was rotated around a circle using the rotation device, and the output voltage values corresponding to different angles were recorded. Finally, the results are normalized and expressed in logarithmic form to obtain the directivity curve of the PVH at this frequency. The rotational speed of the rotary device is 120°/min, and the recording interval of the recorder is 0.2 s, which means that the angle interval for the directivity test is 0.4°. As can be seen from Figure 15b, the results show that the PVH has a good and smooth "8"-shape cosine directivity at 800 Hz. The depths of the sensitivity notch for the X and Y channels are 32.46 dB and 32.08 dB, respectively. The depth of the sensitivity notch for each channel is greater than 32 dB, which meets the application requirements.

5. Conclusions

In this study, a piezoelectric MEMS accelerometer was designed and fabricated for vector hydrophones with four-cantilever beam integrated inertial mass unit structure manufactured using MEMS technology. The cantilever beam of PMA is comprised of Si, SiO_2, Pt/Ti, PZT, and Au, arranged from bottom to top. The high dielectric constant PZT film was employed to measure the transverse sensitivity of the d_{31} working mode. The accelerometer was tested on a vibration platform. The results indicate that the resonant frequency of the PMA is 2300 Hz, with high anti-interference ability, good response linearity, and passive stability. The accelerometer exhibits a perfect response to the input signals, indicating high sensitivity of 33.65 mV/g at 600 Hz. Additionally, the PMA was encapsulated inside the polyurethane waterproof cap of the vector hydrophone. The PVH was tested in a standing wave barrel, revealing excellent low-frequency characteristics, with a receiving sensitivity of −178.99 dB@1000 Hz (0 dB = 1 V/μPa), which is 11 dB higher than the bionic cilia MEMS vector hydrophone (CVH). All channels of PVH exhibited good "8" shape directivity. The fabricated piezoelectric MEMS accelerometer has excellent performance and manufacturing feasibility, indicating its potential for widespread use in various fields requiring high-sensitivity acceleration measurement and underwater target detection applications. The future work will focus on investigating the PVH performance in low-temperature deep sea environments.

Author Contributions: S.S. and L.M. contributed equally to this work. Conceptualization, S.S. and L.M.; data curation, H.M. and K.K.; formal analysis, S.S. and L.M.; funding acquisition, Y.P. and H.M.; investigation, J.Z. and J.H.; methodology, S.S., Y.P. and Z.W.; resources, J.H.; software, H.M. and K.K.; supervision, Z.W.; validation, L.M. and Z.W.; writing—original draft, S.S. and J.Z.; writing—review and editing, S.S., K.K. and Y.P. All authors have read and agreed to the published version of the manuscript.

Funding: This work is supported by the doctoral research start-up fund of Hebei University of Architecture [grant number: B-202301]; Basic Scientific Research Business Project of Colleges and Universities in Hebei Province [grant number: 2022QNJS01]; Hebei Province "333 Talent Project" Funded Project under Grant [grant number: C20221029]; Young Top Talent Project of Hebei Provincial Department of Education [grant number: BJK2023116]; Zhangjiakou Basic Research and Talent Training Program Project [grant numbers: 2311005A, 2311006A]; Zhangjiakou City key research and development plan special [grant number: 2311022B].

Data Availability Statement: Not applicable.

Conflicts of Interest: The authors declare no conflict of interest.

References

1. Saheban, H.; Kordrostami, Z. Hydrophones, fundamental features, design considerations, and various structures: A review. *Sens. Actuators A Phys.* **2021**, *329*, 112790. [CrossRef]
2. Abdul, B.; Mastronardi, V.M.; Qualtieri, A.; Algieri, L.; Guido, F.; Rizzi, F.; De Vittorio, M. Sensitivity and Directivity Analysis of Piezoelectric Ultrasonic Cantilever-Based MEMS Hydrophone for Underwater Applications. *J. Mar. Sci. Eng.* **2020**, *8*, 748. [CrossRef]
3. Ozevin, D. MEMS Acoustic Emission Sensors. *Appl. Sci.* **2020**, *10*, 8966. [CrossRef]

4. Zhang, G.; Liu, M.; Shen, N.; Wang, X.; Zhang, W. The development of the differential MEMS vector hydrophone. *Sensors* **2017**, *17*, 1332. [CrossRef] [PubMed]

5. Wang, R.; Shen, W.; Zhang, W.; Song, J.; Zhang, W. Design and implementation of a jellyfish otolith-inspired MEMS vector hydrophone for low-frequency detection. *Microsyst. Nanoeng.* **2021**, *7*, 1. [CrossRef]

6. Sun, Y.; Yu, Y.; Zhang, C.; Ma, J.; Zheng, Y. Single-sensor pressure-gradient piezoelectric cylindrical tube vector hydrophone. *Vibroeng. Procedia* **2022**, *41*, 124–130. [CrossRef]

7. Xue, C.; Chen, S.; Zhang, W.; Zhang, B.; Zhang, G.; Qiao, H. Design, fabrication, and preliminary characterization of a novel MEMS bionic vector hydrophone. *Microelectron. J.* **2007**, *38*, 1021–1026. [CrossRef]

8. Mireles, J., Jr.; Sauceda, Á.; Jiménez, A.; Ramos, M.; Gonzalez-Landaeta, R. Design and Development of a MOEMS Accelerometer Using SOI Technology. *Micromachines* **2023**, *14*, 231. [CrossRef] [PubMed]

9. Khimunin, A.S. The maximum possible hydrophone size in the studies of the ultrasonic field structure. *Acoust. Phys.* **2008**, *54*, 279–287. [CrossRef]

10. Zhu, J.; Liu, X.; Shi, Q.; He, T.; Sun, Z.; Guo, X.; Liu, W.; Sulaiman, O.B.; Dong, B.; Lee, C. Development trends and perspectives of future sensors and MEMS/NEMS. *Micromachines* **2019**, *11*, 7. [CrossRef]

11. Song, J.; He, C.; Wang, R.; Xue, Y.; Zhang, W. A mathematical model of a piezoresistive eight-beam three-axis accelerometer with simulation and experimental validation. *Sensors* **2018**, *18*, 3641. [CrossRef] [PubMed]

12. Gupta, N.; Dutta, S.; Panchal, A.; Yadav, I.; Kumar, S.; Parmar, Y. Design and fabrication of SOI technology based MEMS differential capacitive accelerometer structure. *J. Mater. Sci. Mater. Electron.* **2019**, *30*, 15705–15714. [CrossRef]

13. Li, S.; Zhao, X.; Bai, Y.; Li, Y.; Ai, C.; Wen, D. Fabrication technology and characteristics research of the acceleration sensor based on Li-doped ZnO piezoelectric thin films. *Micromachines* **2018**, *9*, 178. [CrossRef]

14. Cai, S.; Li, W.; Zou, H.; Bao, H.; Zhang, K.; Wang, J.; Song, Z.; Li, X. Design, Fabrication, and Testing of a Monolithically Integrated Tri-Axis High-Shock Accelerometer in Single (111)-Silicon Wafer. *Micromachines* **2019**, *10*, 227. [CrossRef]

15. Benmessaoud, M.; Nasreddine, M.M. Optimization of MEMS capacitive accelerometer. *Microsyst. Technol.* **2013**, *19*, 713–720. [CrossRef]

16. Kanda, K.; Aiba, T.; Maenaka, K. Piezoelectric MEMS Energy Harvester from Airflow at Low Flow Velocities. *Sens. Mater.* **2022**, *34*, 1879–1888. [CrossRef]

17. Trolier-McKinstry, S.; Muralt, P. Thin Film Piezoelectrics for MEMS. *J. Electroceram.* **2004**, *12*, 7–17. [CrossRef]

18. Ding, X.; Wu, Z.; Gao, M.; Chen, M.; Li, J.; Wu, T.; Lou, L. A High-Sensitivity Bowel Sound Electronic Monitor Based on Piezoelectric Micromachined Ultrasonic Transducers. *Micromachines* **2022**, *13*, 2221. [CrossRef]

19. Hindrichsen, C.C.; Thomsen, E.V.; Lou-Moller, R.; Bove, T. MEMS accelerometer with screen printed piezoelectric thick film. In Proceedings of the 2006 IEEE Sensors, Daegu, Republic of Korea, 22–25 October 2006.

20. Hindrichsen, C.C.; Almind, N.S.; Brodersen, S.H.; Hansen, O.; Thomsen, E.V. Analytical model of a PZT thick-film triaxial accelerometer for optimum design. *IEEE Sens. J.* **2009**, *9*, 419–429. [CrossRef]

21. Hindrichsen, C.C.; Almind, N.S.; Brodersen, S.H.; Lou-Møller, R.; Hansen, K.; Thomsen, E.V. Triaxial MEMS accelerometer with screen printed PZT thick film. *J. Electroceram.* **2010**, *25*, 108–115. [CrossRef]

22. Xu, M.H.; Zhou, H.; Zhu, L.H.; Shen, J.N.; Guo, H. Design and fabrication of a d33-mode piezoelectric micro-accelerometer. *Microsyst. Technol.* **2019**, *25*, 4465–4474. [CrossRef]

23. Lee, Y.C.; Tsai, C.C.; Li, C.Y.; Liou, Y.C.; Chu, S.Y. Fabrication and function examination of PZT-based MEMS accelerometers. *Ceram. Int.* **2021**, *47*, 24458–24465. [CrossRef]

24. Kim, J.; Pyo, S.; Roh, Y. Analysis of a thickness-shear mode vibrator for the accelerometer in vector hydrophones. *Sens. Actuators A Phys.* **2017**, *266*, 9–14. [CrossRef]

25. Pyo, S.; Kim, J.; Kim, H.; Roh, Y. Development of vector hydrophone using thickness–shear mode piezoelectric single crystal accelerometer. *Sens. Actuators A Phys.* **2018**, *283*, 220–227. [CrossRef]

26. Cho, Y.; Je, Y.; Jeong, W.B. A miniaturized acoustic vector sensor with PIN-PMN-PT single crystal cantilever beam accelerometers. *Acta Acust.* **2020**, *4*, 5. [CrossRef]

27. Roh, T.; Yeo, H.G.; Joh, C.; Roh, Y.; Kim, K.; Seo, H.-s.; Choi, H. Fabrication and Underwater Testing of a Vector Hydrophone Comprising a Triaxial Piezoelectric Accelerometer and Spherical Hydrophone. *Sensors* **2022**, *22*, 9796. [CrossRef]

28. Kim, S.B.; Park, H.; Kim, S.H.; Wikle, H.C.; Park, J.H.; Kim, D.J. Comparison of MEMS PZT Cantilevers Based on d31 and d33 Modes for Vibration Energy Harvesting. *Microelectromechan. Syst.* **2012**, *22*, 26–33. [CrossRef]

29. Liu, H.C.; Zhong, J.W.; Lee, C.K.; Lee, S.-W.; Lin, L.W. A comprehensive review on piezoelectric energy harvesting technology: Materials, mechanisms, and applications. *Appl. Phys. Rev.* **2018**, *5*, 041306. [CrossRef]

30. Li, H.D.; Tian, C.; Deng, Z.D. Energy harvesting from low frequency applications using piezoelectric materials. *Appl. Phys. Rev.* **2014**, *1*, 041301. [CrossRef]

31. Zhang, Z.; Zhang, L.; Wu, Z.; Gao, Y.; Lou, L. A High-Sensitivity MEMS Accelerometer Using a $Sc_{0.8}Al_{0.2}N$-Based Four Beam Structure. *Micromachines* **2023**, *14*, 1069.

32. Anton, S.R.; Sodono, H.A. A review of power harvesting using piezoelectric materials. *Smart Mater. Struct.* **2007**, *16*, R1–R21. [CrossRef]

33. Wang, R.; Liu, Y.; Xu, W.; Bai, B.; Zhang, G.; Liu, J.; Xiong, J.; Zhang, W.; Xue, C.; Zhang, B. "Fitness-Wheel-Shaped" MEMS Vector Hydrophone for 3D Spatial Acoustic Orientation. *J. Micromech. Microeng.* **2017**, *27*, 045015. [CrossRef]

34. Wei, H.; Geng, W.; Bi, K.; Li, T.; Li, X.; Qiao, X.; Shi, Y.; Zhang, H.; Zhao, C.; Xue, G.; et al. High-Performance Piezoelectric-Type MEMS Vibration Sensor Based on LiNbO3 Single-Crystal Cantilever Beams. *Micromachines* **2022**, *13*, 329. [CrossRef] [PubMed]
35. Yang, X.; Xu, Q.; Zhang, G.; Zhang, L.; Wang, W.; Shang, Z.; Shi, Y.; Li, C.; Wang, R.; Zhang, W. Design and implementation of hollow cilium cylinder MEMS vector hydrophone. *Measurement* **2021**, *168*, 108309. [CrossRef]
36. Zhao, A.B.; Zhou, B.; Li, J.Q. Calibration of vector hydrophones in deep water. *Appl. Mech. Mater.* **2013**, *336*, 303–308. [CrossRef]

micromachines

MDPI

Article

Structural Design of MEMS Acceleration Sensor Based on PZT Plate Capacitance Detection

Min Cui [1,*], Senhui Chuai [2], Yong Huang [3], Yang Liu [2] and Jian Li [4]

[1] Shanxi Key Laboratory of Information Detection and Processing, North University of China, Taiyuan 030051, China
[2] School of Instrumentation and Electronics, North University of China, Taiyuan 030051, China; sz202206204@st.nuc.edu.cn (S.C.); liuyang03042022@163.com (Y.L.)
[3] Shanghai Institute of Aerospace Control Technology, Shanghai 201109, China; huangyong9608@163.com
[4] School of Information and Communication Engineering, North University of China, Taiyuan 030051, China; lijian@nuc.edu.cn
* Correspondence: cmcm_1980930@163.com

Abstract: The problem that the fuze overload signal sticks and is not easily identified by the counting layer during the high-speed intrusion of the projectile is an important factor affecting the explosion of the projectile in the specified layer. A three-pole plate dual-capacitance acceleration sensor based on the capacitive sensor principle is constructed in this paper. The modal simulation of the sensor structure is carried out using COMSOL 6.1 simulation software, the structural parameters of the sensor are derived from the mechanical properties of the model, and finally the physical sensor is processed and fabricated using the derived structural parameters. The mechanical impact characteristics of the model under different overloads were investigated using ANSYS/LS-DYNA, and the numerical simulation of the projectile intrusion into the three-layer concrete slab was carried out using LS-DYNA. Under different overload conditions, the sensor was tested using the Machette's hammer test and the output signal of the sensor was obtained. The output signal was analyzed. Finally, a sensor with self-powered output, high output voltage amplitude, and low spurious interference was obtained. The results show that the ceramic capacitive sensor has a reasonable structure, can reliably receive vibration signals, and has certain engineering applications in the intrusion meter layer.

Keywords: PZT; layer counting identification; capacitive acceleration sensor; MEMS; LS-DYNA

Citation: Cui, M.; Chuai, S.; Huang, Y.; Liu, Y.; Li, J. Structural Design of MEMS Acceleration Sensor Based on PZT Plate Capacitance Detection. *Micromachines* **2023**, *14*, 1565. https://doi.org/10.3390/mi14081565

Academic Editor: Ha Duong Ngo

Received: 15 July 2023
Revised: 4 August 2023
Accepted: 4 August 2023
Published: 6 August 2023

1. Introduction

In modern war, in order to prevent important targets from being destroyed by the enemy, targets of strategic significance (underground command center, missile silo, weapons depot, underground communication hub, etc.) are often hidden underground with extremely strong fortifications [1]. For example, the Japanese Central Command Post, which was completed and put into use in 1984, has two floors above ground and three underground, with a total depth of more than 30 m [2]. The Command and Operation Center of the French Staff Headquarters is a two-story and half-story building composed of steel bars and high-strength concrete, with the lowest ground level at -10 m [3]. The underground command center of the Iraqi Presidential Palace is 18 m above the ground, and its main structure is made of special high-strength and heat-resistant concrete with a thickness of up to two meters [4]. Therefore, how to effectively attack this kind of target has become the focus for various countries and military departments [5]. Various research institutions are constant committed to finding more effective means and methods of damage [6]. Based on this goal, penetration ammunition was born. Penetration ammunition mainly relies on the kinetic energy of the warhead to invade the inside of the target; after penetrating the target, the intelligent fuze to controls the location of the projectile blast point to achieve the maximum damage effect [7]. Penetration munitions can be used in various weapon systems [8], including air force weapons such as aerial bombs and air-to-ground missiles, lethal weapons such as strategic guided missiles and ground-to-ground missiles, and naval weapons such as ship-borne cruise missiles, ship-to-ground missiles, and so on [9]. In general, penetration munitions are aimed at high-value, hardened targets and are therefore mostly used in a variety of precision strike weapons [10]. According to the types of targets, penetration munitions can be used to target important ground targets such as airport runways, bridges, ground command centers, and communication hubs. They can also effectively strike various underground command centers, missile silos, weapons depots, and underground communication hubs [11]. They can also strike a variety of maritime targets, including the precision killing of aircraft carrier fleets, frigates, and reef military bunkers [12].

Due to the in-depth research on high-speed and ultrahigh-speed intrusion munitions technology at home and abroad, the multi-layer target intrusion layer counting problem of high-speed large aspect ratio projectiles has become a hot topic of research in the field of intrusion at home and abroad [13]. The MEMS island-beam acceleration sensor has the technical advantage of a high range [14], and its layer counting sensing technology is usually applied to the multi-layer target layer counting of high-speed and large aspect ratio projectiles (initial velocity of intrusion greater than 600 m/s). However, the following problems still exist:

When the aspect ratio of the projectile increases and the projectile velocity is greater than 800 m/s, the stress wave propagates back and forth in the projectile body during the penetration process, resulting in concussive acceleration [15]. The peak value of the high-frequency concussive signal will increase rapidly and the overload envelope of multi-layer target penetration will be completely submerged, resulting in the mutual adhesion of overload signals from layer to layer, which cannot effectively identify the penetration layer [16].

The problem of signal adhesion is difficult to solve at the root. The maximum destructive effect of an intrusive munition must be achieved by intelligent fuze control and detonation at the right location [17]. A fuze is a special single-use product defined as a control system that can use information about the target, environment, platform, and network to detonate or ignite the combatant charge according to a predetermined strategy. It can select the point of detonation and give instructions for range or extended engine ignition, and information about the destructive effect [18]. There are four commonly used fuze initiation methods for intrusion munitions: timed initiation, layer/cavity initiation, stroke initiation, and media recognition initiation [19]. Among the four initiation methods, the layer/cavity counting initiation is the most widely studied and the most intensively researched initiation method. The principle of layer/cavity counting identification fuze is that when the strike target is a multi-layer hard target, the acceleration sensor in the fuze receives the acceleration signal output penetrated each hard target layer, so that the electrical signal system inside the fuze processes the acceleration signal (the size of the deceleration, derivative, inflection point, etc.); and analyze the number of layers of the target that have penetrated the target to drive the position of the projectile. When the projectile reaches a predetermined number of target layers, the fuze emits a detonation signal and detonates the explosive to achieve the optimal destruction of the penetrated target [20]. The existing fuze layer counting techniques often rely on single-axis acceleration sensors to sense the overload of the projectile as it penetrates each layer of the target and the calculation of the number of target layers is achieved by using a comparison of the overload signal amplitude and threshold; great progress has been made in engineering applications. However, the complexity of the penetration problem, including the oscillation of the projectile, the friction between the projectile and the target plate, and the characteristics of the sensor, can lead to a large amount of clutter interference in the overload signal, and in some cases the overload signal amplitude of the penetrated target is not obvious, leading to difficulties in fuze layer counting or even layer counting errors Therefore, in order to further improve the accuracy of the fuze layer, it is necessary to explore better sensors to solve the interference problem at the source of the overload signal generation and to study more accurate and reliable fuze layer counting strategies. In this paper, we focus on the precise identification of multilayer targets for target penetration munition fuzes. In the layer counting identification of an intruder fuze, the main layer counting method currently relies on high g-value acceleration sensors or an acceleration threshold switch to realize layer counting; however, there are two very important key points to accurately count the number of layers through which the intruder warhead has passed:

(1)   The occurrence of the overload signal
(2)   The way the overload signal is processed and discriminated.

This paper focuses on the precise identification of multilayer targets for intrusion munitions fuzes. A flat plate capacitive acceleration sensor based on a three-layer pole plate is designed for the layer counting recognition of intrusion munition fuzes. The corresponding simulation and tests are conducted to obtain a sensor with excellent output performance.

2. Design Principle

2.1. Three-Dimensional Acceleration Numerical Simulation of Projectile Penetration into a Multilayer Concrete Target

With the continuous development of modern science, especially the development of computer technology and various intrusion theories, numerical simulation by computers has become a very critical and important technology [21]. The penetration problem is a very complex problem, involving the calculation of the deformation of the bullet and the target plate, material stress, target plate erosion, and a series of key points. The previous empirical and analytical methods have certain shortcomings [22]. The empirical method requires a large number of live firing tests and analysis of the measured data, which requires huge manpower and material resources and is less economical [23]. The analytical method requires the establishment of certain prerequisites, and the scope of application is more restricted [24]. The numerical method relies on electronic computers and combines the concept of finite elements or finite volumes to achieve the purpose of studying engineering and physical problems and even various problems in nature by means of numerical calculations and image display [25].

In order to accurately verify the rationality and usability of the designed sensor, LS-DYNA 19.2 finite element simulation software was used to produce and solve the whole process of projectile intrusion into the concrete target plate and obtain the overload signal of the smart fuze inside the projectile during the projectile intrusion, which supports the accurate layer counting identification of the subsequent fuze [26].

Firstly, the LS-DYNA finite element simulation software was used to simulate the dynamics of the projectile body penetrating the three-layered cement board.

When designing the projectile penetration model, the initial velocity of the projectile was set to 800 m/s, the thickness of the reinforced concrete target plate was 40 mm, the spacing of the target plate was 300 mm, 400 mm, or 500 mm, the diameter of the projectile body was 60 mm, and the vertical length was 200 mm; the penetration model is shown in Figure 1.

The velocity and acceleration changes of the projectile penetration simulation are shown in Figure 2. Figure 2a,b show the changes in projectile velocity and acceleration during the penetration process. According to the images, when the projectile completely penetrates the first target plate, the horizontal velocity decreases from 800 m/s to 724.39 m/s; the velocity difference is 75.61 m/s and the maximum acceleration is −257,176.87 g. When the projectile penetrates the second target plate, the horizontal velocity decreases to 659.3 m/s, the velocity difference is about 65.09 m/s, and the maximum acceleration is −189,766.764 g. When the projectile completely penetrates the third target plate, the horizontal velocity decreases to 619.31 m/s, the velocity difference is about 39.99 m/s, and the maximum acceleration is −116,588.921 g.

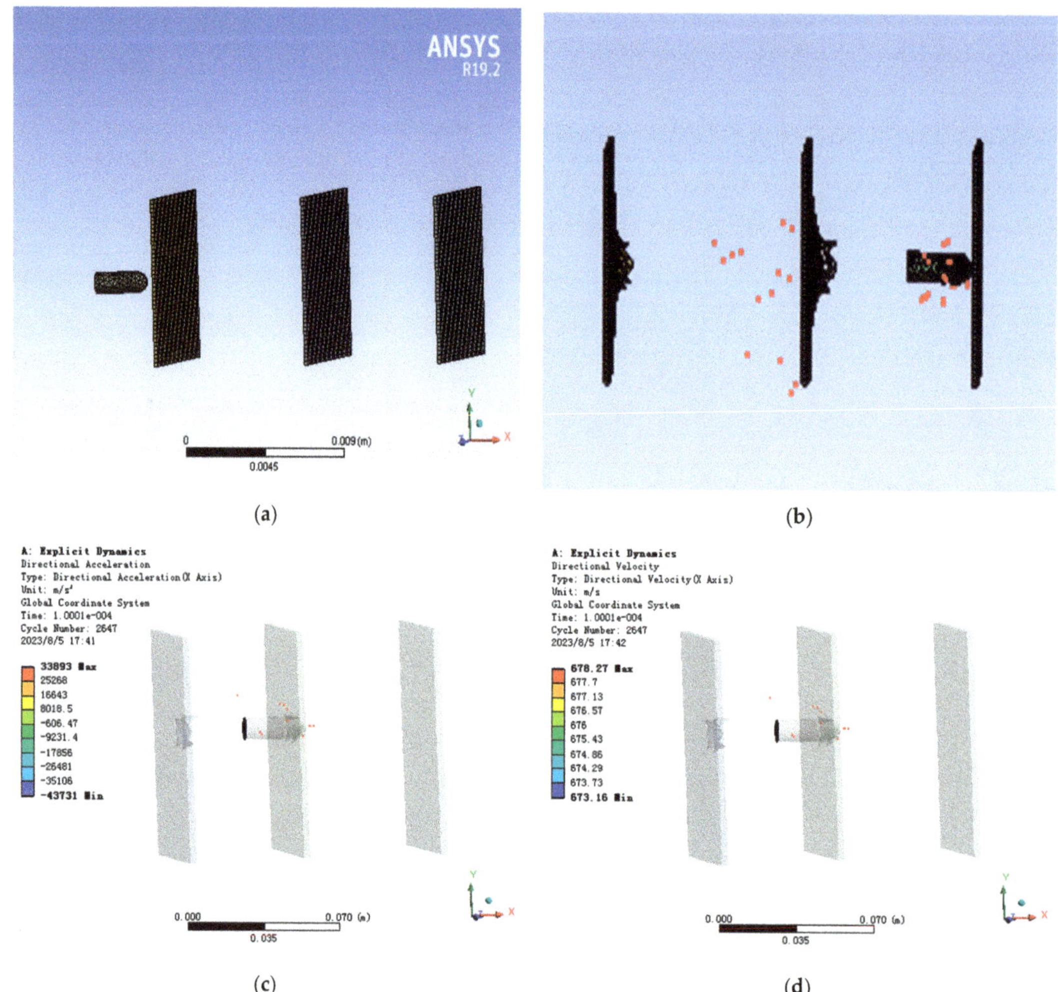

Figure 1. Simulation diagram of projectile penetration of three-layer plate: (**a**) is the physical model of the projectile penetrating the three-layer target, (**b**) is the penetration diagram when the acceleration is whole, (**c**) is the penetration diagram of the tail velocity, and (**d**) is the penetration diagram of the tail acceleration.

We place the probe at the rear of the projectile to detect the velocity and acceleration of the projectile during penetration. The probe diagram is shown in Figure 1c,d. Figure 2c,d show the velocity and acceleration curves at the rear of the projectile. From Figure 2d, it can be seen that when the projectile penetrates the cement plate, due to the internal structure of the projectile, the maximum acceleration caused by the penetration resistance propagating to the rear of the projectile is about 10,000 g.

The core of this layer counting principle approach lies in the extraction of acceleration peaks and the setting of the thresholds. Using the intrusion into multiple layers of targets, the acceleration will suddenly change to produce a peak and the counting of the number of layers penetrating the target is achieved by comparing the magnitude of this peak with a set threshold [27].

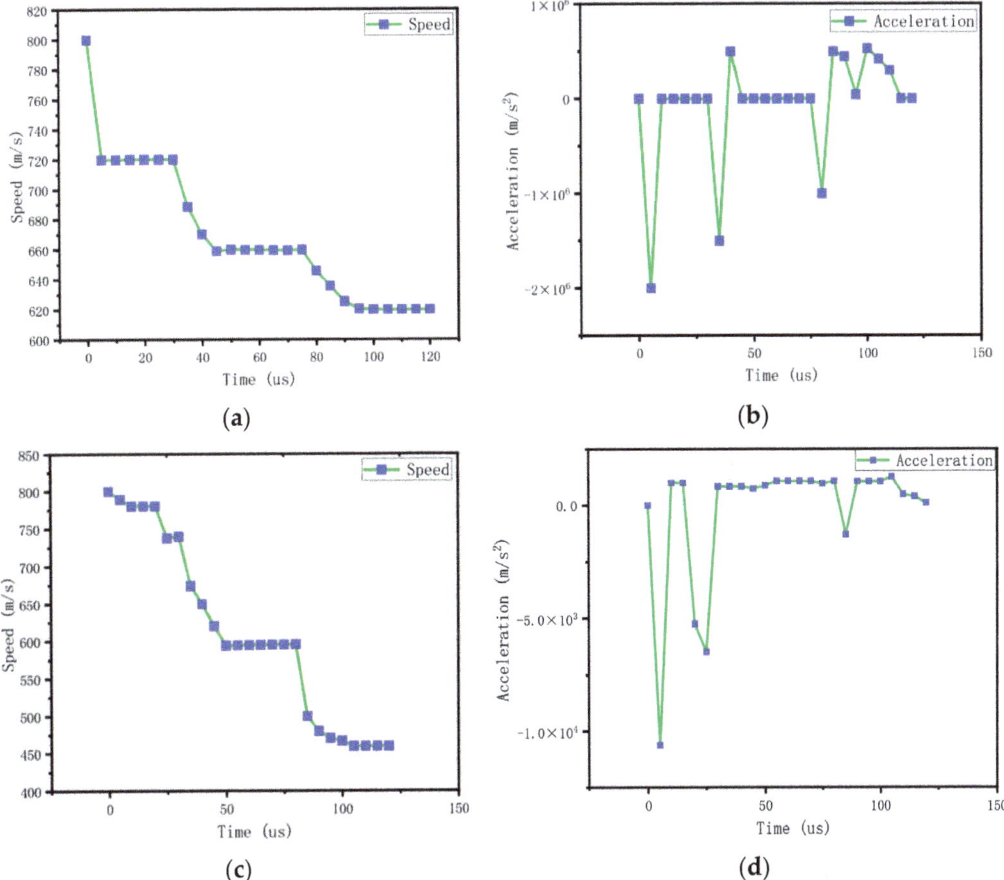

Figure 2. The overall velocity and acceleration curves of the projectile, as well as the velocity and acceleration curves of the projectile: (**a**) is the velocity of the projectile as a whole, (**b**) is the acceleration of the projectile as a whole, (**c**) is the velocity of the projectile tail, and (**d**) is the acceleration of the projectile tail.

It can be seen that the projectile took about 120 μs from the beginning of contacting the first target plate to leaving the last target plate and that the projectile decreased by a total of 180 m/s during the process of penetrating the three target plates. Overall, the acceleration overload values of the three layers of penetration were large and the acceleration sensors could easily obtain their corresponding acceleration signals.

2.2. Principle of Capacitance Sensor Measurement

The sensor element of a capacitive sensor is various types of capacitors. When the external measurement changes, it leads to the change in capacitor capacitance. Through the corresponding external measurement circuit, the capacitance change of a capacitor is converted into an electrical signal through the corresponding mathematical relationship. Then, by measuring the size of electrical signals, the changes and sizes measured can be judged [28].

The capacitive acceleration sensor designed in this paper has a double-capacitance structure. The overall structure is cylindrical, consisting of movable upper and middle plates, a fixed lower plate, and upper and lower zirconia ceramic dielectric layers, the diameter of the plates and dielectric layers is 4.5 mm, the thickness of the three plates is 0.5 mm, and the thickness of the ceramic dielectric layer is 1.5 mm. Its structure is shown in Figure 3a, and the equivalent dynamics model is shown in Figure 3b.

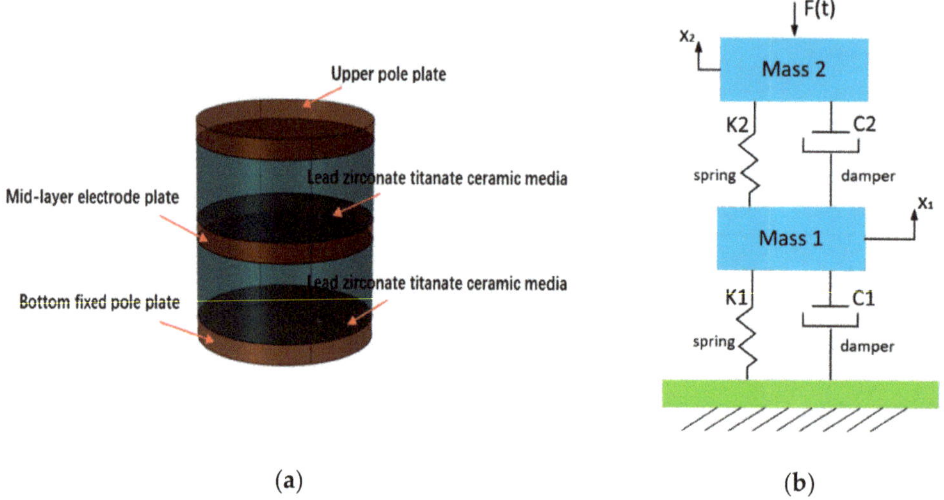

Figure 3. Overall structure of double-plate capacitive accelerometer and equivalent dynamics model of the double-plate capacitive accelerometer. (**a**) Overall structure of the three-electrode plate capacitance accelerometer and (**b**) equivalent kinetic model.

The traditional acceleration sensor often adopts a single cantilever beam, a double cantilever beam, or piezoresistive structure [29]. The sensor internal sensitive element also has the property of material damping after deformation, but due to its own structural reasons, the material damping area is small, resulting in the shock attenuation of the sensor after the impact; if the shock signal lasts a long time, the sensor cannot quickly restore to the initial state [30]. Based on the above reasons, the structure of the sensor was changed, the damping area of the material was increased, and the damping of the sensor was increased so that the vibration of the sensor can be rapidly attenuated. The structure can effectively improve the damping coefficient of the system, reduce the time of shock response, and eliminate the problem of signal adhesion [31]. The movable upper plate, middle plate, and fixed lower plate are all made of palladium silver (Pd-Ag) conductor material, and the ceramic dielectric layer is made of zirconia ceramic material. Due to the excellent characteristics of zero mechanical hysteretic, high elasticity, corrosion resistance, wear resistance, creep, small hysteretic, and high temperature robustness of the zirconia ceramic sheet, Hooke's law is strictly followed until cracking. In addition, the strength of the material during compression is much higher than that during tension [32].

Generally speaking, the input acceleration range of a capacitive sensor structure with a circular plate is greater than that of a capacitive sensor structure with a square plate of the same size [31]. Therefore, the electrode plate and dielectric layer are designed to be circular [33]. The common operating principles of capacitive sensors include variable pole pitch, variable area, and a variable dielectric constant [34]. Considering the high impact force of projectile penetration, the capacitive sensor is designed as a variable-polar-pitch type using a high-strength, impact-resistant [35], and hard dielectric material, as shown in Figure 3.

In the three-pole plate capacitive sensor, the dielectric vibrates during the intrusion process and, at this moment, when the dielectric moves between the plates, damping is generated between the two plates. This damping includes the damping generated when the dielectric vibrates and the air damping caused by some air left in the gap during the manufacturing process. Damping is a destructive factor in maintaining vibrations. In many cases, measures are taken to reduce the damping so that vibration can be maintained at a minimum energy complement per cycle. However, in many other cases, damping is deliberately introduced into the system to reduce oscillations. A notable example is the micro-accelerometer. Damping is necessary, and proper damping should be considered from the design stage.

As shown in Figure 3b, since the lower pole plate is fixed, the upper pole plate and the middle pole plate can be equivalently regarded as masses m_1 and m_2. The stress–strain is transferred between each two pole plates through the dielectric. The electrode plate generates stress–strain through its own elastic deformation, and at the same time, there will be a damping effect. Similarly, the dielectric will have elastic deformation and damping effect in the process of transferring stress. Therefore, the elastic deformation and damping between the electrode plate and the dielectric are equivalent to the springer and the dampers, which constitute the dynamics model of the mass-spring-damping system. The kinetic equation of its kinetic model can be expressed as:

$$\begin{bmatrix} m_1 & 0 \\ 0 & m_2 \end{bmatrix}\begin{bmatrix} \ddot{x}_1 \\ \ddot{x}_2 \end{bmatrix} + \begin{bmatrix} c_1 + c_2 & -c_2 \\ -c_2 & c_2 \end{bmatrix}\begin{bmatrix} \dot{x}_1 \\ \dot{x}_2 \end{bmatrix} + \begin{bmatrix} k_1 + k_2 & -k_2 \\ -k_2 & k_2 \end{bmatrix}\begin{bmatrix} x_1 \\ x_2 \end{bmatrix} = \begin{bmatrix} 0 \\ F(t) \end{bmatrix} \tag{1}$$

where m_1, m_2, k_1, k_2, c_1, c_2, x_1 and x_2 are the mass, stiffness, equivalent damping coefficient, and displacement of the middle and upper pole plates, respectively, and where $F(t) = (m_1 + m_2)\, a_1$, a_1 is the impact acceleration experienced by the sensor during penetration. The differential equation of Equation (1) can be transformed into a binary equation by Laplace variation as follows:

$$\begin{cases} m_1 s^2 x_1(s) + (c_1 + c_2)s x_1(s) - c_2 s x_2(s) + (k_1 + k_2)x_1(s) - k_2 x_2(s) = 0 \\ m_2 s^2 x_2(s) - c_2 s x_1(s) + c_2 s x_2(s) - k_2 x_1(s) + k_2 x_2(s) = F(s) \end{cases} \tag{2}$$

Then, $s = j\omega$ is substituted into Equation (2) and the vibration amplitude of the upper and middle plates is obtained by complex operation and joint solution:

$$x_1 = \frac{cF(t)}{ab - c^2}, x_2 = \frac{bF(t)}{ab - c^2} \tag{3}$$

In the formula:

$$a = k_2 - m_2\omega^2 + \omega c_2 j;\ b = k_1 + k_2 - m_1\omega^2 + \omega(c_1 + c_2)j;$$

$$c = k_2 + \omega c_2 j;\ \omega = \sqrt{\frac{k_2}{m_2}};$$

The total capacitance of the sensor consists of two capacitance components formed by the three pole plates of the sensor:

$$C = C_1 + C_2 \tag{4}$$

$$C_1 = \frac{\varepsilon_0 \varepsilon_r A}{d_0 - x_1}, C_2 = \frac{\varepsilon_0 \varepsilon_r A}{d_0 - x_2} \tag{5}$$

$$C_1 = \frac{\varepsilon_0 \varepsilon_r A}{d_0 - x_1} = \frac{\varepsilon_0 \varepsilon_r A}{d_0\left(1 - \frac{x_1}{d_0}\right)} = \frac{\varepsilon_0 \varepsilon_r A\left(1 + \frac{x_1}{d_0}\right)}{d_0\left(1 - \frac{x_1^2}{d_0^2}\right)}$$

$$C_2 = \frac{\varepsilon_0 \varepsilon_r A}{d_0 - x_2} = \frac{\varepsilon_0 \varepsilon_r A}{d_0\left(1 - \frac{x_2}{d_0}\right)} = \frac{\varepsilon_0 \varepsilon_r A\left(1 + \frac{x_2}{d_0}\right)}{d_0\left(1 - \frac{x_2^2}{d_0^2}\right)} \tag{6}$$

Since the vibration amplitude is much smaller than the initial pole–plate spacing, Equation (6) can be reduced to the following equation:

$$C_1 = C_0 - C_0 \frac{x_1}{d_0}, C_2 = C_0 - C_0 \frac{x_2}{d_0} \tag{7}$$

Bringing Equation (3) into Equation (7) yields the relationship between acceleration and capacitance:

$$C = C_0 - C_0\left(\frac{(b+c)(m_1+m_2)a_1}{d_0(ab-c^2)}\right) \tag{8}$$

3. COMSOL Finite Element Simulation

3.1. COMSOL Mode Simulation

The first four orders of modal simulation of the sensor structure were carried out using COMSOL software, and the simulation results are shown in Table 1. The simulation results show that the capacitor has an intrinsic frequency of 80.739 kHz in the direction of pulse acceleration in the first-order mode. The specific fourth-order modal diagram is shown in Figure 4

Table 1. Fourth-mode frequency.

Modal Order	Vibration Frequency (kHz)
1	80.739
2	113.631
3	128.231
4	128.234

The frequency response cutoff frequency caused by the penetration impact is less than 20 kHz, and the natural frequency of the three-layer capacitor structure should be 3–5 times higher than the cutoff frequency of the target signal. This is to ensure the accuracy of the dynamic measurement. The first-order natural frequency of the three-layer plate capacitor sensor is 4 times the cutoff frequency of the target signal; therefore, the double-layer plate capacitor can meet the requirements of no measurement distortion.

3.2. Impact Simulation of Double-Layer Capacitive Accelerometer

In the process of projectile penetration, the triple electrode plate acceleration sensor will be subjected to a high impact, so the impact displacement simulation of the sensor plate is carried out to determine whether the impact displacement of the plate conforms to the theory of small deflection deformation.

The response frequency during projectile penetration is about 20 kHz. The impact on the sensor during projectile penetration into the concrete slab is simulated by giving the capacitive accelerometer (a) 40,000 g/40 µs, (b) 100,000 g/40 µs, (c) 150,000 g/40 µs, and (d) 200,000 g/40 µs. This is shown in Figure 5. The maximum displacement at the top pole plate under the impact of 200,000 g/40 µs is 2.23 microns, which is consistent with the theory of small deflection deformation.

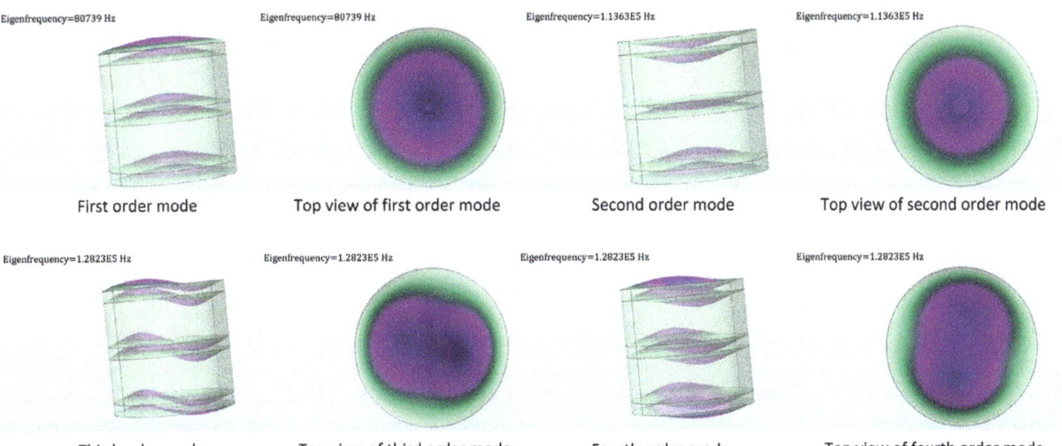

Figure 4. Fourth mode diagram of double-layer plate capacitive accelerometer.

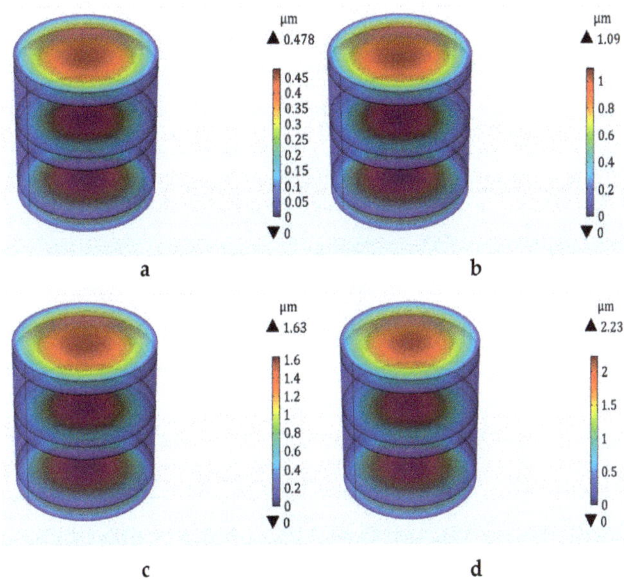

Figure 5. Shock displacement plots of double capacitance accelerometers under different shocks. (**a**) 40,000 g impact displacement map, (**b**) 100,000 g impact displacement map, (**c**) 150,000 g impact displacement map, (**d**) 200,000 g impact displacement map.

The sensor is affected by the penetrating stress wave at the tail of the projectile, and the impact acceleration of the sensor is about 10,000 g. Therefore, we conducted a 10,000 g impact analysis on the sensor and observed the maximum displacement and stress of the sensor. Through the simulation, we found that under the sinusoidal impact of 10,000 g/40 µs, the displacement of the sensor was 0.11 micron. The maximum stress was 0.136 MPa. It can be seen that the structure was relatively stable at a 10,000 g impact. We learn from Figure 2b that the impact of the entire projectile body upon penetration can reach 200,000 g.

Therefore, we carried out impact simulation on the sensor over the range, and the result was that the maximum displacement of the plate was 2.23 microns and the maximum stress was 68 MPa. It can be considered that the sensor has a stable structure under a 200,000 g impact and can still output linearly. The 200,000 g impact is the penetration impact size that the projectile as a whole can achieve. We use the maximum penetration impact to judge whether the sensor can stabilize output. Through simulation, we can see that the structure can withstand 200,000 g of impact without damage.

As shown in Figure 1, we conducted penetration simulation on the projectile, and the projectile carried out penetration motion at the initial velocity of 800 m/s. The acceleration of the projectile was with the process of projectile penetration into the cement plate and change. The specific value changes, as shown in Figure 2. Because the fuze is located at the tail of the projectile, when the peak impact of the penetration to the rear of the projectile is about 10,000–15,000 g, the natural frequency of the sensor is 80.739 kHz. The impact frequency caused by the impact of each layer of the projectile is about 20 kHz, that is, the time to penetrate each layer of the pole plate is 40 microseconds.

Giving a peak acceleration of 10,000 g/120 μs to a three-layer electrode plate capacitive sensor, the equivalent displacement of the upper pole plate surface is shown in Figure 6. When the projectile penetrates the first layer of cement plate, the velocity is maximum and the displacement is at its maximum at this time; the cement plate will have a resistance to the projectile and the projectile will produce a reverse acceleration accordingly. The upper pole plate displacement gradually decreases as the acceleration decreases. In the second and third layers of the target, the peak displacement of the upper pole plate is smaller than that in the first layer of the target.

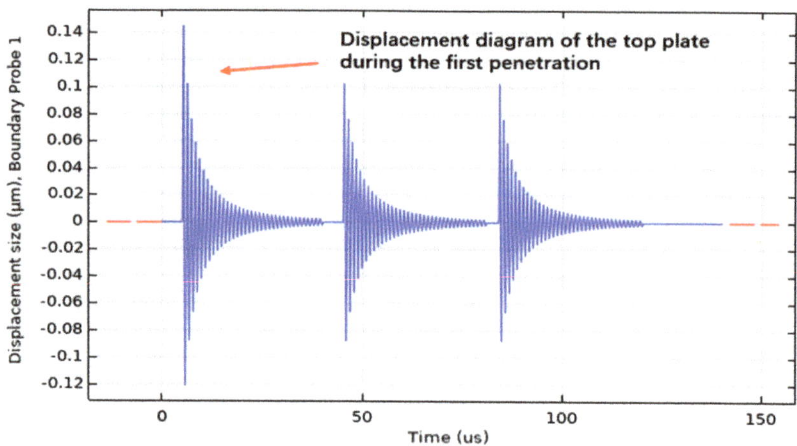

Figure 6. Sensor displacement curve.

When the projectile penetrates the first layer of cement plate, the velocity is the maximum, at this time, the displacement is the maximum; when the acceleration decreases, the displacement of the upper pole plate gradually decreases.

4. Process Design of the Plate Capacitive Accelerometer

The main process steps of the double-plate capacitor are shown in Figure 7 [36]. This is unlike [36], where we used lead zirconate titanate as a dielectric. The specific steps are as follows: (1) Firstly, two pieces of lead zirconate titanate piezoelectric ceramics (PZT) are machined to the designed dimensions by mechanical processing, as shown in Figure 7a. (2) Using the thick film process, the silver conductor paste was screen printed on the two ceramics by screen printing, as shown in Figure 7b. (3) Dry film lamination was then performed to form a photosensitive etched impedance layer to form a capacitive plate, as shown in Figure 7c. (4) The ceramic sheet was ground to form a complete double-layer plate capacitor structure, which was pre-sintered under an infrared light lamp for half an hour at room temperature to prevent the formation of structural bubbles. The plate molecules were then recrystallized in a tunnel furnace to improve the strength of the capacitor structure. Finally, it was cooled in a vacuum for 1.5 h. The results are shown in Figure 7d,e. (5) The upper and lower pole plates were brushed with silver conductor paste for dry film lamination, as shown in Figure 7f. (6) The prepared double-layer flat plate capacitor was taken out and the three flat plates were soldered together with leads to make a capacitive sensing element, as shown in Figure 7g.

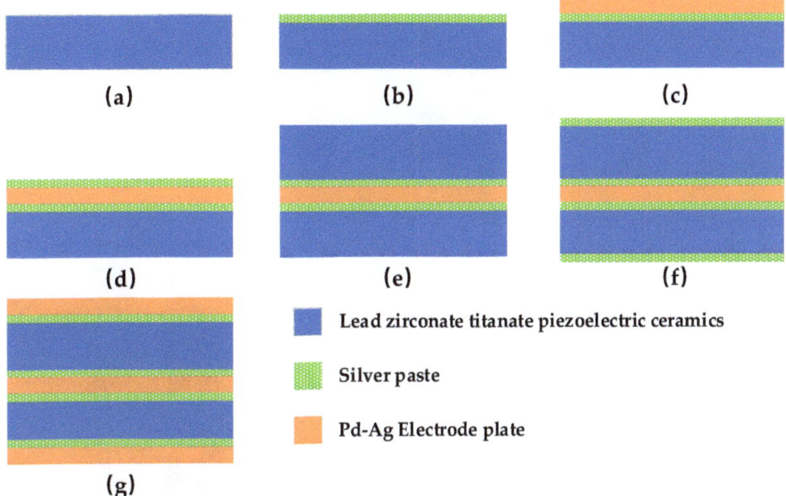

Figure 7. Schematic diagram of process steps. (**a**) PZT electrode plate. (**b**) Screen printing (**c**) Dry film lamination (**d**) Double-layer plate capacitor (**e**) Sintered two PZT plates (**f**) Top screen printing (**g**) Electrode plate welding.

5. Experiment and Analysis

5.1. Machette Hammer Impact Experiment

According to the simulation, the maximum impact on the projectile tail where the sensor is located during penetration is about 10,000 g, so a Machette's hammer test was used for impact testing.

The test is an impact Machette shock test on the designed double-pole plate capacitor. The Machette hammer consists of four parts: the hammer head (fixture and sensor), the teeth, the weight, and the anvil. The test instrument is shown in Figure 8.

A certain acceleration is determined with a Machette hammer to simulate the force acting on the internal acceleration sensor when the projectile penetrates the concrete target. The effective measurement range of the sensor is 0–200,000 g. According to the simulation test, the target material is constructed with C30 concrete. When the initial penetration velocity is 800 m/s, the peak overload acceleration of the projectile body is usually in the range of 10,000–20,000 g. The maximum acceleration pulse width of the Machette hammer used was 120 μs, and the maximum impact acceleration was 50,000 g. Therefore, this experiment can simulate the penetration process of single-layer target.

After the sensor test platform was built, the Machette hammer was used to test the electrical performance of the sensor.

Impact the sensor with a Machette hammer at 12,520 g. The output image of the sensor voltage is shown in Figure 9.

Figure 8. Machette's hammer test equipment.

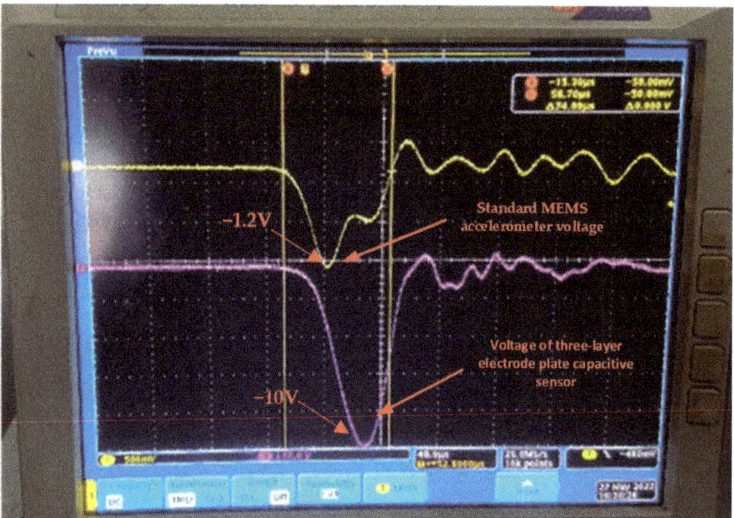

Figure 9. Voltage output diagram of the sensor.

5.2. Data Analysis

From the purple waveform in Figure 8, it can be seen that the test experiment results in the signal change trend and simulation results remain consistent and the voltage signal output and the sensor changes are similar; in crossing a layer of cement board, the signal is also first sharply increased until it reaches a peak voltage of −10 V; the size of the peak voltage is also the sensor of the overload shock accurately perceived by the embodiment. After reaching the peak, the voltage falls back quickly until it reaches zero voltage. At this point, the plates squeeze each other and bounce back and the sensor voltage signal increases again; however, because the strength of the plate bounce back after the collision is small, and coupled with the material damping that exists from beginning to end and the continuous attenuation of the movement of the plates, the amplitude of the bounce back is small and the sensor bounce output voltage signal is also small, about a quarter of the peak voltage. The yellow signal in the figure is the output signal of the MEMS accelerometer, and the purple line represents the output response of the three-layer plate acceleration sensor. The MEMS accelerometer was compared with an Endevco accelerometer. With its high sensitivity and fast response time, the Endevco accelerometer is widely used in many fields, such as mechanical engineering, the automotive industry, aerospace, and medical equipment. It has great sensitivity and stability and is able to provide accurate measurement results over a wide range of temperatures and frequencies. At the same time, it is well designed to minimize noise interference and provide a clear acceleration signal. Additionally, the Endevco accelerometer has a range of 0–100,000 g and a sensitivity of 0.00001 V/g.

As can be seen in Figure 9, the response amplitude of the three-layer electrode plate capacitance sensor is large, the fallback is free of noise, and the response is superior to the output signals of the Endevco accelerometer used for the comparison.

6. Conclusions

In this paper, a dual-capacitance acceleration sensor based on three-layer flat plate is designed. Aiming at the problem of excessive clutter interference for the fuze overload signal in the process of target penetration, a theoretical model of the capacitive sensor is constructed from the perspective of a sensor by using the principle of a capacitive sensor. From the perspective of material, zirconia was selected as the dielectric material and a palladium–silver alloy as the plate material. The structure of double-layer plate capacitor was designed. The transient and modal simulation of the sensor is carried out by COMSOL simulation software, and the influence law of the structural parameters on the sensor performance is obtained and the appropriate parameter values are determined. Moreover, the three-dimensional overload finite element simulation analysis model of the full-size projectile penetrating the three-layer concrete target plate is established by LS-DYNA and the velocity acceleration of the projectile penetrating is analyzed. The manufacturing process of the double-plate capacitor is introduced. The sensor is processed and manufactured and the output characteristics of the sensor are tested using a Machette hammer. Finally, a capacitive acceleration sensor with stable output voltage and less clutter is obtained.

Author Contributions: Y.L. and S.C. conceived and designed the experiments; Y.H. and Y.L. performed the experiments; S.C. analyzed the data; J.L. organized the data for the paper and provided the testbed. Supervision of the experiments was provided; M.C. wrote the paper. All authors have read and agreed to the published version of the manuscript.

Funding: This work is supported by Technology Field Fund of Basic Strengthening Plan of China (2021-JCJQ-JJ-0720).

Data Availability Statement: Not applicable.

Conflicts of Interest: The authors declare no conflict of interest.

References

1. Cui, R.; Li, K.; Xu, X.; Shen, C.; Shi, Y.; Cao, H. Design and experiment of MEMS solid-state wave gyroscope quadrature error correction system. *IEEE Sens. J.* **2023**, *23*, 16645–16655. [CrossRef]
2. Ning, J.; Ren, H.; Li, Z.; Xu, X. A mass abrasion model with the melting and cutting mechanisms during high-speed projectile penetration into concrete slabs. *Acta Mech. Sin.* **2022**, *38*, 121597. [CrossRef]
3. Tusnio, J. An Electronic Time Fuze with an Overload-Time Protection Device. *Prz. Elektrotech.* **2010**, *86*, 214–217.
4. Peng, Y.; Wu, H.; Fang, Q.; Gong, Z. Deceleration time of projectile penetration/perforation into a concrete target: Experiment and discussions. *Adv. Struct. Eng.* **2019**, *22*, 112–125. [CrossRef]
5. Tais, A.S.; Ibraheem, O.F.; Raoof, S.M. Effect of Thickness and Reinforcement on Concrete Plates under High Speed Projectiles. *Struct. Eng. Mech.* **2022**, *82*, 587–594. [CrossRef]
6. Gerami, N.D.; Khazraiyan, N. Evaluation of semi-analytical model for rigid projectile penetration into concrete/metal targets. *Struct. Concr.* **2020**, *21*, 117–128. [CrossRef]
7. Feldgun, V.; Yankelevsky, D.; Karinski, Y. New method for predicting perforation parameters of a concrete slab target to a penetrating projectile. *Int. J. Impact Eng.* **2023**, *174*, 104512. [CrossRef]

8. Xu, L.-Y.; Cai, F.; Xue, Y.-Y.; Takahashi, C.; Li, Y.-Y. Numerical Analyses of Local Damage of Concrete Slabs by Normal Impact of Deformable Solid Projectiles. *KSCE J. Civ. Eng.* **2019**, *23*, 5121–5132. [CrossRef]
9. Meng, Q. Numerical Simulation of Multifunctional Projectile Penetrating Reinforced Concrete Target Plate Based on Sensor Data Acquisition. *J. Sens.* **2022**, *2022*, 3115123. [CrossRef]
10. Wang, Z.-L.; Li, Y.-C.; Shen, R.; Wang, J. Numerical study on craters and penetration of concrete slab by ogive-nose steel projectile. *Comput. Geotech.* **2007**, *34*, 1–9. [CrossRef]
11. Ye, S.; Xu, Y.; Zhou, Y.; Cheng, J.; Huang, J.; Cai, Y.; Yao, X.; Luo, S. Penetration dynamics of steel spheres into a ballistic gelatin: Experiments, nondimensional analysis, and finite element modeling. *Int. J. Impact Eng.* **2022**, *162*, 104144. [CrossRef]
12. Deng, J.; Zhang, X.; Liu, C.; Wang, W. Penetration performance of axisymmetric U-shape-nose grooved projectile into aluminum target: Theoretical model and experiment. *Lat. Am. J. Solids Struct.* **2018**, *15*. [CrossRef]
13. Kong, X.Z.; Wu, H.; Fang, Q.; Zhang, W.; Xiao, Y.K. Projectile penetration into mortar targets with a broad range of striking velocities: Test and analyses. *Int. J. Impact Eng.* **2017**, *106*, 18–29. [CrossRef]
14. Wang, R.; Tang, E.; Yang, G.; Gao, G.; Wang, L. Research on layer-counting experimental simulation system for projectile penetrating multi-layered targets. *Measurement* **2020**, *151*, 107108. [CrossRef]
15. Koh, H.S.; Shin, D.Y.; Yoon, G.H. The role of granular buoyant force of projectile in determining the penetration depth. *Int. J. Impact Eng.* **2022**, *166*, 104238. [CrossRef]
16. Furubayashi, Y.; Oshima, T.; Yamawaki, T.; Watanabe, K.; Mori, K.; Mori, N.; Matsumoto, A.; Kamada, Y.; Isobe, A.; Sekiguchi, T. A 22-Ng/Root Hz 17-MW Capacitive MEMS Accelerometer with Electrically Separated Mass Structure and Digital Noise-Reduction Techniques. *IEEE J. Solid-State Circuits* **2020**, *55*, 2539–2552. [CrossRef]
17. Shi, Y.; Zhang, J.; Jiao, J.; Zhao, R.; Cao, H. Calibration Analysis of High-G MEMS Accelerometer Sensor Based on Wavelet and Wavelet Packet Denoising. *Sensors* **2021**, *21*, 1231. [CrossRef]
18. Li, X.; Hu, J.; Liu, X. A High-Performance Digital Interface Circuit for a High-Q Micro-Electromechanical System Accelerometer. *Micromachines* **2018**, *9*, 675. [CrossRef]
19. Zhang, J.; Shi, Y.; Cao, H.; Zhao, S.; Zhao, Y.; Wang, Y.; Zhao, R.; Hou, X.; He, J.; Chou, X. Design and Implementation of a Novel Membrane-Island Structured MEMS Accelerometer with an Ultra-High Range. *IEEE Sens. J.* **2022**, *22*, 20246–20256. [CrossRef]
20. Cai, P.; Xiong, X.; Wang, K.; Wang, J.; Zou, X. An Improved Difference Temperature Compensation Method for MEMS Resonant Accelerometers. *Micromachines* **2021**, *12*, 1022. [CrossRef]
21. Shi, Y.; Zhao, Y.; Feng, H.; Cao, H.; Tang, J.; Li, J.; Zhao, R.; Liu, J. Design, fabrication and calibration of a high-G MEMS accelerometer. *Sens. Actuators A Phys.* **2018**, *279*, 733–742. [CrossRef]
22. Shi, Y.; Wang, Y.; Feng, H.; Zhao, R.; Cao, H.; Liu, J. Design, Fabrication and Test of a Low Range Capacitive Accelerometer with Anti-Overload Characteristics. *IEEE Access* **2020**, *8*, 26085–26093. [CrossRef]
23. Guo, C.; Shi, Y.; Cao, H.; Wen, X.; Zhao, R. Dynamic parameter identification of a high g accelerometer based on BP-PSO algorithm. *Sens. Actuators A Phys.* **2023**, *349*, 114024. [CrossRef]
24. Duan, C.; Jiang, Y.; Chen, L.; Tai, H.; He, Y. Design and Development of MEMS Capacitive Large-Scale Strain Sensors. *Integr. Ferroelectr.* **2013**, *147*, 123–130. [CrossRef]
25. Almutairi, B.; Alshehri, A.; Kraft, M. Design and Implementation of a MASH2-0 Electromechanical Sigma–Delta Modulator for Capacitive MEMS Sensors Using Dual Quantization Method. *J. Microelectromech. Syst.* **2015**, *24*, 1251–1263. [CrossRef]
26. Lu, Q.; Shen, C.; Cao, H.; Shi, Y.; Liu, J. Fusion Algorithm-Based Temperature Compensation Method for High-G MEMS Accelerometer. *Shock. Vib.* **2019**, *2019*, 3154845. [CrossRef]
27. Lu, Q.; Pang, L.; Huang, H.; Shen, C.; Cao, H.; Shi, Y.; Liu, J. High-G Calibration Denoising Method for High-G MEMS Accelerometer Based on EMD and Wavelet Threshold. *Micromachines* **2019**, *10*, 134. [CrossRef]
28. Huang, L.; Yang, H.; Gao, Y.; Zhao, L.; Liang, J. Design and Implementation of a Micromechanical Silicon Resonant Accelerometer. *Sensors* **2013**, *13*, 15785–15804. [CrossRef]
29. Huang, C.-Y.; Chen, J.-H. Development of Dual-Axis MEMS Accelerometers for Machine Tools Vibration Monitoring. *Appl. Sci.* **2016**, *6*, 201. [CrossRef]
30. Benedetti, E.; Ravanelli, R.; Moroni, M.; Nascetti, A.; Crespi, M. Exploiting Performance of Different Low-Cost Sensors for Small Amplitude Oscillatory Motion Monitoring: Preliminary Comparisons in View of Possible Integration. *J. Sens.* **2016**, *2016*, 7490870. [CrossRef]
31. Kolli, V.R.; Dudla, P.; Talabattula, S. Integrated optical MEMS serially coupled double racetrack resonator based accelerometer. *Optik* **2021**, *236*, 166583. [CrossRef]
32. Shi, Y.; Wen, X.; Zhao, Y.; Zhao, R.; Cao, H.; Liu, J. Investigation and experiment of high shock packaging technology for High-G MEMS accelerometer. *IEEE Sens. J.* **2020**, *20*, 9029–9037. [CrossRef]
33. Yan, Z.; Hou, B.; Zhang, J.; Shen, C.; Shi, Y.; Tang, J.; Cao, H.; Liu, J. MEMS Accelerometer Calibration Denoising Method for Hopkinson Bar System Based on LMD-SE-TFPF. *IEEE Access* **2019**, *7*, 113901–113915. [CrossRef]
34. Langfelder, G.; Bestetti, M.; Gadola, M. Silicon MEMS inertial sensors evolution over a quarter century. *J. Micromech. Microeng.* **2021**, *31*, 084002. [CrossRef]

35. Zhu, L.; Fu, Y.; Chow, R.; Spencer, B.F., Jr.; Park, J.W.; Mechitov, K. Development of a High-Sensitivity Wireless Accelerometer for Structural Health Monitoring. *Sensors* **2018**, *18*, 262. [CrossRef] [PubMed]
36. Cui, M.; Huang, Y.; Li, J.; Meng, M. Design and Experiment of a Novel High-Impact MEMS Ceramic Sandwich Accelerometer for Multi-Layer Target Penetration. *IEEE Access* **2020**, *8*, 107387–107398. [CrossRef]

Article

Experimental Research of Triple Inertial Navigation System Shearer Positioning

Cheng Lu [1,2], Shibo Wang [2,*], Kyoosik Shin [3], Wenbin Dong [3] and Wenqi Li [3]

[1] School of Mechanical Engineering, Anhui Science and Technology University, Chuzhou 233100, China; luch@ahstu.edu.cn

[2] School of Mechanical and Electrical Engineering, China University of Mining & Technology, Xuzhou 221116, China

[3] Department of Mechatronics Engineering, Hanyang University, Ansan 15588, Gyeonggi-do, Republic of Korea; kss_1212@163.com (K.S.); wuhui0514@163.com (W.D.); liwenqi@hanyang.ac.kr (W.L.)

* Correspondence: wangshb@cumt.edu.cn

Abstract: In order to improve the positioning accuracy of shearers, the overground experimental device based on the positioning model of TINS (Triple Inertial Navigation System) was built. The influence of TINS installation parameters on positioning accuracy was discussed through two sets of experiments: the inter-INS (Inertial Navigation System) distances influence experiments and the tri-INS plane spatial position influence experiments. The results show that the positioning accuracy of the shearer is improved to a different extent under the two sets of experimental conditions. When the inter-INS distances are 0.2 m, the positioning accuracy is the highest and the positioning accuracy improvement effect is also the best. When the negative plane α_3 is 45°, the positioning accuracy is the highest, and the positioning accuracy improvement effect is also the best. The analysis shows that the main factor affecting the positioning accuracy is the precision of the evaluated values outputs of TINS from EKF (Extended Kalman Filter). Considering the positioning accuracy, equipment installation convenience and so on, the optimum installation parameters are 90° (horizontal installation) α_3 for the positive plane and 0.2 m inter-INS distances.

Keywords: shearer; TINS; multi-INS; positioning method; coal mine-intelligent

Citation: Lu, C.; Wang, S.; Shin, K.; Dong, W.; Li, W. Experimental Research of Triple Inertial Navigation System Shearer Positioning. *Micromachines* **2023**, *14*, 1474. https://doi.org/10.3390/mi14071474

Academic Editors: Huiliang Cao and Niall Tait

Received: 9 May 2023
Revised: 13 July 2023
Accepted: 20 July 2023
Published: 23 July 2023

1. Introduction

Intelligent and unmanned mining offers the prospect of safe and efficient mining [1]. The key to solving the core problem of the current development of intelligent mining is the real-time and accurate positioning of the shearer in the seam space of coal [2].

The INS positioning of shearers is an independent positioning technology that has been widely recognized by scientific research institutions and coal mining equipment manufacturers [3]. Although it has many technical advantages, it lacks the ability to eliminate and reduce the accumulated errors of inertial sensors over time. As a result, the method of improving INS positioning accuracy has become a research hotspot all over the world. As early as 2001, Australia's CSIRO (Commonwealth Scientific and Industrial Research Organization) began to study the integrated inertial navigation system of shearers, which was a Doppler velocity measurement radar with Zero-Speed Correction Technology [4,5]. In the later study, the scholar Ralston [6] from the aforementioned organization designed a kind of Kalman filter algorithm to optimize the positioning accuracy of a shearer based on the closed paths formed by the shearer's moving trajectory. Chinese scholars have also undertaken a variety of research on improving INS positioning accuracy of the shearer. From 2016 to now, the scientific research team of the authors has proposed Shearer Positioning Technology Combined with INS and Coder [7], Shearer INS Positioning Dynamic Zero-Speed Correction Technology [8], Dynamic Precise Positioning Technology Based on Closed Path Optimal Estimation Model [9], Dual Inertial Navigation Positioning Technology of Shearer [10], Multiple Inertial Navigation Redundancy Positioning Technology of Shearer [11], Shearer Positioning Technology Based on Heterogeneous Multi-source Information Fusion [12], Inertial Navigation and Ultra-broadband Fusion Positioning Technology of Shearer [13,14], Accuracy Improvement Method Based on Shearer Kinematic Constraint [15], Two-Point Method Based Deviation Compensation Algorithm Based on Two Points [2], and so on. The measurement error and installation deviation of INS have been investigated systematically.

Although previous scholars have conducted research on various methods to improve the INS positioning accuracy of shearers, they have mainly focused on aspects such as combining INS with other sensors to form an integrated navigation system or exploiting the kinematic constraints. The analysis of the reasons affecting the INS positioning accuracy mainly focused on the installation errors or inertial sensor errors [6,16,17]. Few scholars have conducted their research on multi-INS positioning technology of shearers and analysis of reasons influencing multi-INS positioning accuracy. However, the research of multi-INS are widespread in the aviation industry and other fields. Liu Zhi [18] took the differences of attitude angles of two INS as the observed values and took the zero velocity errors of accelerometers as the estimated values to realize the dynamic correction of accelerometer errors. Zhang Linlin [19] corrected the velocity error to some extent by using two or more INS damping networks of ships. To improve the INS accuracy of civil aircraft, Bai Junqiang [20] proposed an optimal navigation solution with three parallel INS. Through theoretical analysis and simulations, they concluded that the positioning accuracy could be improved under both normal operating conditions and fault conditions. Si [21] integrated the main INS positioning information and the INS positioning information of airborne weapons of fighter aircraft, thus improving the positioning accuracy of the weapon.

It can be seen from the above references that multi-INS technology is a feasible way of improving positioning accuracy. Compared with satellite, infrared, laser, ultrasonic and other positioning methods, multi-INS technology can improve the positioning accuracy by INS itself, which is more suitable for the complex coal mine environment with no satellite signal and high dust. In this paper, the TINS positioning model of shearers was investigated and the influences of installation parameters on the positioning accuracy were analyzed through experiments. In addition, the optimum installation parameters of TINS were proposed considering various factors. The experimental conclusions can provide some guidance for the application of multi-INS technology in the field of shearer positioning.

2. The TINS Positioning Principle of Shearers

The installation of the three INS on the shearer is shown in Figure 1. The geometric center of the upper surface of the shearer is defined as the origin (O_b) of the shearer coordinate system (b system). INS-1 is installed at the origin of the shearer coordinate system, so the position of INS-1 represents the position of shearer. α is the angle between the Z_b axis and the line defined by INS-1 and INS-3. The geographic coordinate system (ENU, East-North-Up) is selected as the navigation coordinate system (n-system). The coder is installed on the axle of the shearer's walking mechanism.

The composition and working mechanism of the TINS is shown in Figure 2. The three INS collect attitude angle values of shearer at the same time, then the transform cosine matrix from b-system to n-system is calculated (as shown in Equation (1)). The coder collects the displacements of the shear in b-system. It is converted to n-system by the transform cosine matrix (as shown in Equation (2)) and accumulated to the initial position of each INS in n-system (as shown in Equation (3)). Thus, the position values of the three INS in n-system are obtained. Those positioning values are used as state values (as shown in Equation (4)) and the inter-INS distances are used as observed values (as shown in Equation (5)). More accurate positioning value outputs are obtained from the EKF. The state equation and observed equation are shown in Equation (6) and Equation (8), respectively.

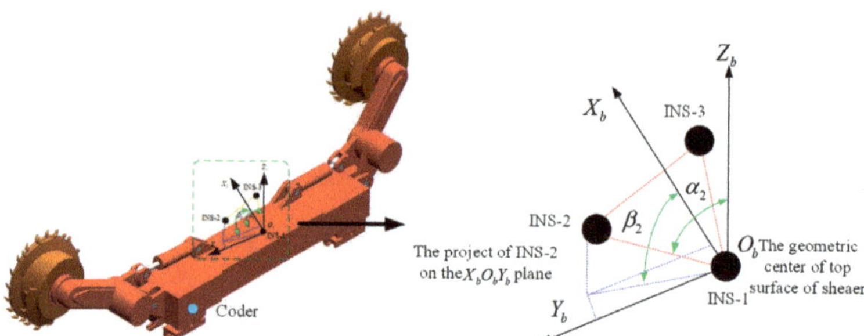

Figure 1. Installation of TINS.

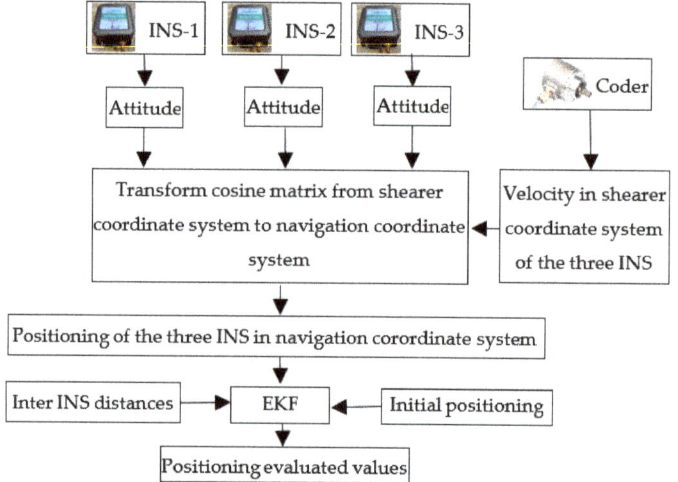

Figure 2. The composition and working mechanism of TINS.

$$C_b^n = \begin{bmatrix} \cos\gamma_i \cos\varphi_i + \sin\gamma_i \sin\theta_i \sin\varphi_i & \cos\theta_i \sin\varphi_i & \sin\gamma_i \cos\varphi_i - \cos\gamma_i \sin\theta_i \sin\varphi_i \\ -\cos\gamma_i \sin\varphi_i + \sin\gamma_i \sin\theta_i \cos\varphi_i & \cos\theta_i \cos\varphi_i & -\sin\gamma_i \sin\varphi_i - \cos\gamma_i \sin\theta_i \cos\varphi_i \\ -\sin\gamma_i \cos\theta_i & \sin\theta_i & \cos\gamma_i \cos\theta_i \end{bmatrix} \tag{1}$$

φ, θ and γ are the heading, pitch and rolling measured by INS-i.

$$\Delta S_i^n = \begin{bmatrix} \cos\theta_i \sin\varphi_i \\ \cos\theta_i \cos\varphi_i \\ \sin\theta_i \end{bmatrix} s \tag{2}$$

φ and θ are the heading and pitch collected by INS-i, ΔS_i^n is the displacement of INS-i in n-system and s is the shearer's displacement collected by the coder.

$$P_i^n(k) = [X_i, Y_i, Z_i]_k^T = P_i^n(0) + \sum_{j=1}^{k} \Delta S_i^n(j) \tag{3}$$

$P_i^n(k)$ is the position of INS-i in n-system at k time. $P_i^n(0)$ is the position of INS-i in n-system at 0 time. X_i, Y_i and Z_i is the coordinate value of north, east and up of INS-i in n-system.

$$X = \begin{pmatrix} x_1 & y_1 & z_1 & x_2 & y_2 & z_2 & x_3 & y_3 & z_3 \end{pmatrix}^T \tag{4}$$

x_i, y_i and z_i ($i = 1, 2, 3$) are the positioning in n-system of INS-i.

$$Z = [r_{12}, r_{13}, r_{23}]^T \tag{5}$$

r_{ij} is the inter-INS distance between INS-i and INS-j.

$$X(k+1) = X(k) + B(k)v(k)T + W(k) \tag{6}$$

$B(k)$ is given by Equation (7), where $v(k)$ is the shearer's instantaneous velocity computed by the coder, T is the sampling period of INS and $W(k)$ is the state noise at time k.

$$B(k) = \begin{bmatrix} \cos\theta_1(k)\sin\varphi_1(k) \\ \cos\theta_1(k)\cos\varphi_1(k) \\ \sin\theta_1(k) \\ \cos\theta_2(k)\sin\varphi_2(k) \\ \cos\theta_2(k)\cos\varphi_2(k) \\ \sin\theta_2(k) \\ \cos\theta_3(k)\sin\varphi_3(k) \\ \cos\theta_3(k)\cos\varphi_3(k) \\ \sin\theta_3(k) \end{bmatrix} \tag{7}$$

r_{ij} is the inter-INS distance between INS-i and INS-j,

$$Z(k) = H \cdot X(k) + V(k) \tag{8}$$

$V(k)$ is the observed noise at time k; H is the Jacobian matrix (shown in Equation (9)),

$$H = \begin{bmatrix} \frac{x_1-x_2}{r_{12}} & \frac{y_1-y_2}{r_{12}} & \frac{z_1-z_2}{r_{12}} & \frac{x_2-x_1}{r_{12}} & \frac{y_2-y_1}{r_{12}} & \frac{z_2-z_1}{r_{12}} & 0 & 0 & 0 \\ \frac{x_1-x_3}{r_{13}} & \frac{y_1-y_3}{r_{13}} & \frac{z_1-z_3}{r_{13}} & 0 & 0 & 0 & \frac{x_3-x_1}{r_{13}} & \frac{y_3-y_1}{r_{13}} & \frac{z_3-z_1}{r_{13}} \\ 0 & 0 & 0 & \frac{x_2-x_3}{r_{23}} & \frac{y_2-y_3}{r_{23}} & \frac{z_2-z_3}{r_{23}} & \frac{x_3-x_2}{r_{23}} & \frac{y_3-y_2}{r_{23}} & \frac{z_3-z_2}{r_{23}} \end{bmatrix} \tag{9}$$

3. Experimental Device and Methods

3.1. Experimental Device

The experimental device is composed of three INS with the same type and a manual turntable, coder, 12 V DC power source, serial ports hub and upper computer. The three INS were manufactured by Beijing CNSENS Technology Co., Ltd. (Beijing, China) and typed IMU680-G. The main performance indexes of their gyroscopes are shown in Table 1. All three INS are connected to serial ports hubs with an RS422 interface. The coder is connected to serial ports hub with an RS485 interface. The two interfaces are converted to a USB interface and connected to the upper computer via a serial ports hub. All the hardware is installed on the mobile vehicle by means of a shelf with three levels (as shown in Figure 3). The relative position of the three INS can be changed by adjusting the height of the second level. The GPS-RTK positioning system is also installed on the third level. The track of the high-precision GPS-RTK mobile station is taken as the real track of the experimental device. It can be seen from the TINS positioning model of the shearer that INS-1 is located at the geometric center of the shearer's upper surface, so it can represent the shearer's positioning value. Therefore, the GPS-RTK mobile station is installed close to INS-1 as a reference of the real position of the experimental device.

Table 1. Main performance indexes of gyroscopes of INS.

Measuring Range	Zero-Bias Stability	Distinguishability	In-Band Noise	Attitude Accuracy
$\pm300°/s$	$\leq18°/h$	$0.03°/s$	$0.3°/s$	<0.3 deg (RMS)

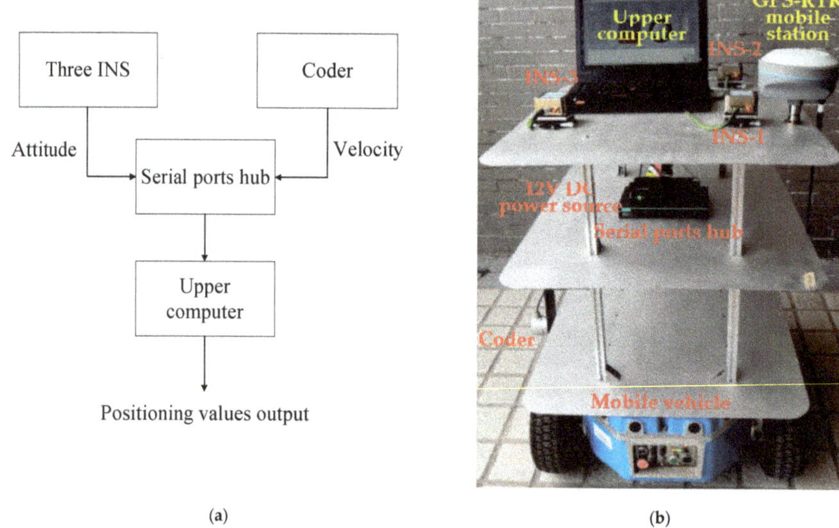

(a) (b)

Figure 3. Experimental device. (**a**) Schematic diagram; (**b**) Hardware.

3.2. Methods

A heading angle of 40° north to east was chosen for ease of operation. The experimental device was moved in a straight line along the gaps between the floor tiles (as shown in Figure 4) to simulate the movement of the shearer along the coal wall. The initial conditions of the experiments were set as follows: the moving speed of the experimental device was set to 1.0 m/s–1.2 m/s, the moving distance of the experimental device was about 60 m and the three INS and the GPS-RTK were started at the same time to synchronize the running time of the two positioning systems. Through the following two sets of experiments, the influence of the inter-INS distances and the spatial position of the tri-INS plane on the positioning accuracy of the shearer was discussed. The experimental procedure is shown in Figure 5. The installation parameters of the experimental device are given in Table 2.

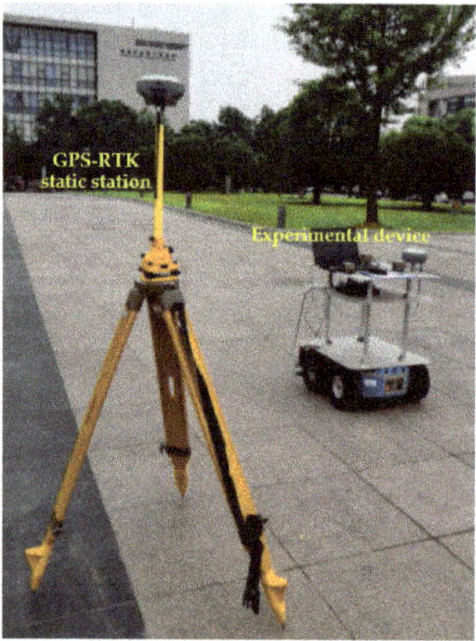

Figure 4. Schematic diagram of the experimental methods.

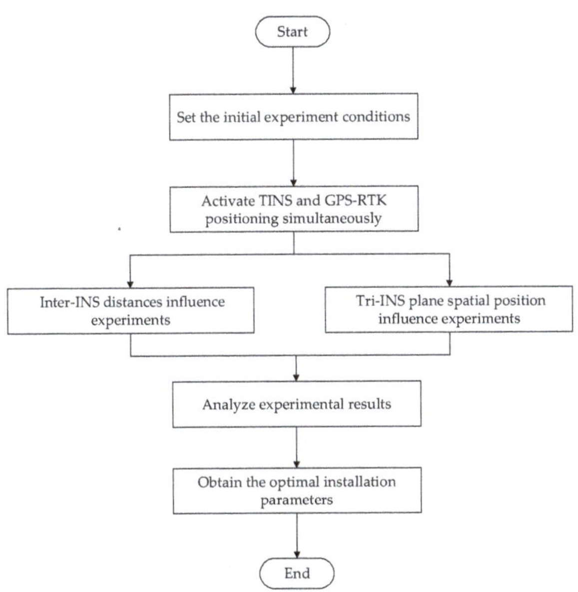

Figure 5. The experimental flow chart.

Table 2. Installation parameters table of experimental device.

| | Inter-INS Distances Influence Experiments | | | | | Tri-INS Plane Spatial Position Influence Experiments | | | | | | | | |
| | | | | | | Positive Plane | | | | | Negative Plane | | | |
No.	1	2	3	4	5	6	7	8	9	10	11	12	13	14
r_{12}	0.1 m	0.2 m	0.28 m	0.38 m	0.38 m	0.38 m	0.38 m	0.38 m	0.38 m	0.38 m	0.38 m	0.38 m	0.38 m	0.38 m
r_{13}	0.1 m	0.2 m	0.28 m	0.38 m	0.38 m	0.38 m	0.38 m	0.38 m	0.38 m	0.38 m	0.38 m	0.38 m	0.38 m	0.38 m
α_3	90°	90°	90°	90°	0°	45°	90°	135°	180°	0°	45°	90°	135°	180°

In order to investigate the influence of the inter-INS distances, the three INS were arranged horizontally, and the inter-INS distances were set as $r_{12} = r_{13} = h$, as shown in Figure 6a. The h values chosen were 0.1 m, 0.2 m, 0.28 m and 0.38 m, respectively.

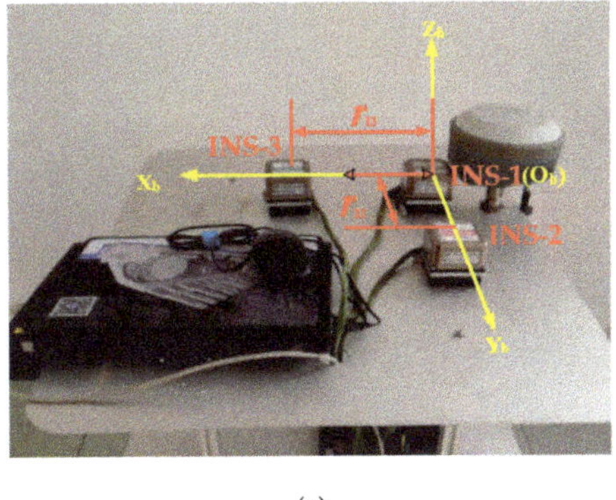

(a)

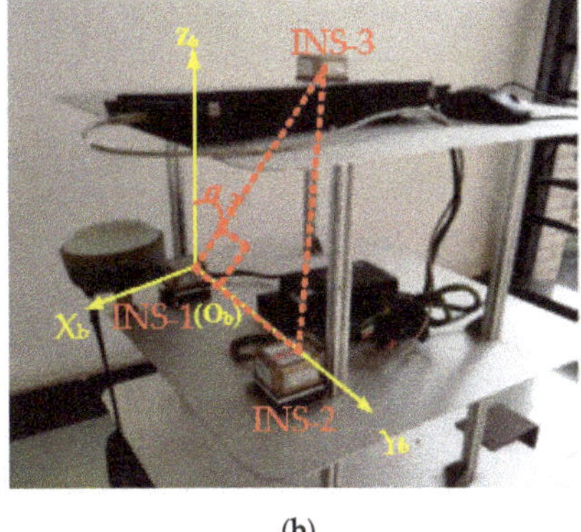

(b)

Figure 6. Schematic diagram of TINS. (**a**) Inter-INS distances influence experiments; (**b**) Tri-INS plane spatial position influence experiments.

To investigate the influence of the Tri-INS plane spatial position, the three INS were arranged at appropriate positions on the second and third layers of the shelf. The height of the second layer was adjusted so that α_3 varied in the range of 0~360° (as shown in Figure 6b), so that the planes determined by the three INS covered the entire three-dimensional space. If the X_b coordinate of INS-3 is positive, it is defined as the positive plane. Otherwise, it is defined as the negative plane.

The Spherical Probability Error (SEP), which is commonly used in the field of navigation and positioning, was used to evaluate the positioning accuracy. The SEP is calculated according to Equation (10).

$$SEP = 0.51(\delta x + \delta y + \delta z) \tag{10}$$

σx, σy and σz are the root mean square of the positioning errors in the east, north and upward directions, respectively.

4. Results and Discussions

4.1. Inter-INS Distances Influence Experiments

Figure 7 shows the TINS positioning trajectories and the single-INS positioning trajectories in the north–east plane and north–up plane of Experiment No.4 in Table 2. As can be seen from the figures, the TINS positioning trajectories are closer to the real trajectories than the single-INS positioning trajectories in the above two planes. The single-INS positioning trajectories fluctuate irregularly in the two planes, and the TINS positioning trajectories curves are relatively smoother.

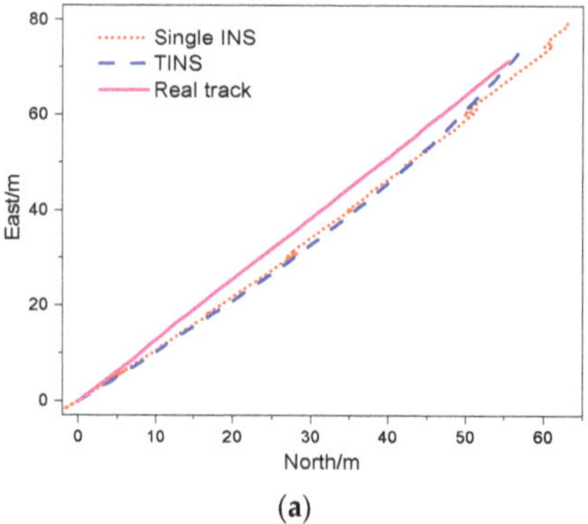

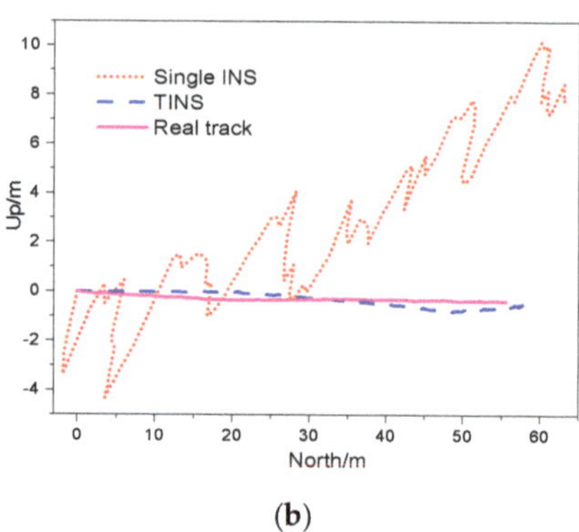

Figure 7. TINS positioning trajectories. (**a**) East-north plane. (**b**) North-up plane.

Figure 8 shows the real-time positioning errors of the inter-INS distances influence experiments in Table 2. The maximum errors of single-INS in east, north and up are 10.29 m, 11.12 m and 10.53 m, respectively, while the maximum errors of TINS are 3.28 m, 3.79 m and 0.44 m, respectively. The maximum error of TINS is obviously lower than that of single-INS. The SEP of single-INS positioning is 7.77 m, and that of TINS positioning is 2.38 m. The accuracy of the TINS is 69.35% better than that of single-INS.

In Experiments No.1, No.2 and No.3 in Table 2, the SEP of single-INS positioning is 6.94 m, 6.67 m and 7.67 m, respectively, while the SEP of TINS positioning is 3.37 m, 1.58 m and 3.38 m (as shown in Figure 9). Compared to the single-INS positioning, the TINS positioning accuracy is improved by 53.77%, 76.31% and 55.93%, respectively. The positioning accuracy of the shearer is improved when the inter-INS distances take different values. The positioning accuracy is the highest and the positioning accuracy improvement effect is also the best when the inter-INS distances are 0.2 m.

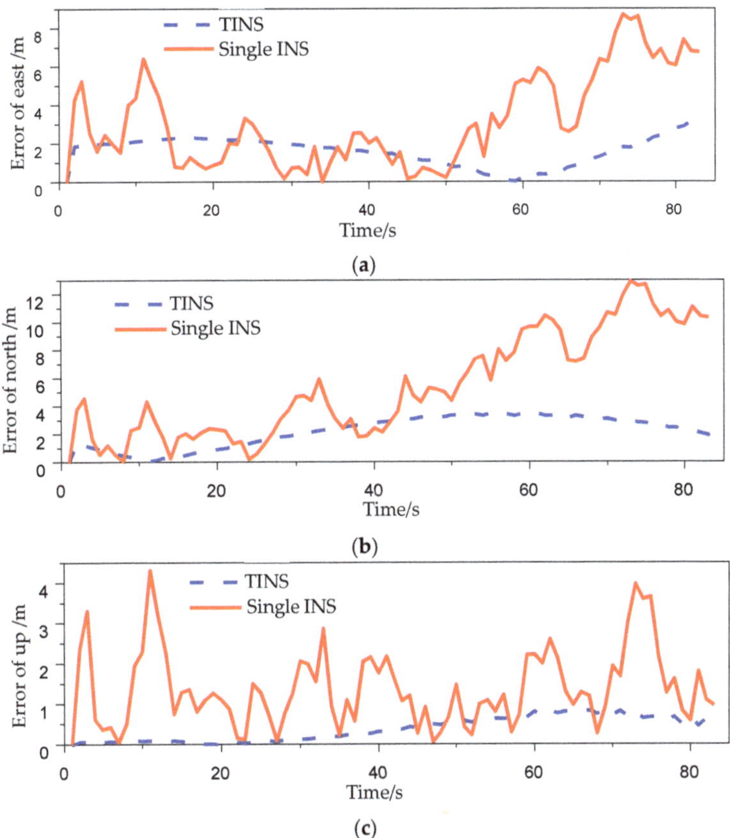

Figure 8. Error of shearer positioning. (**a**) Error of east. (**b**) Error of north. (**c**) Error of up.

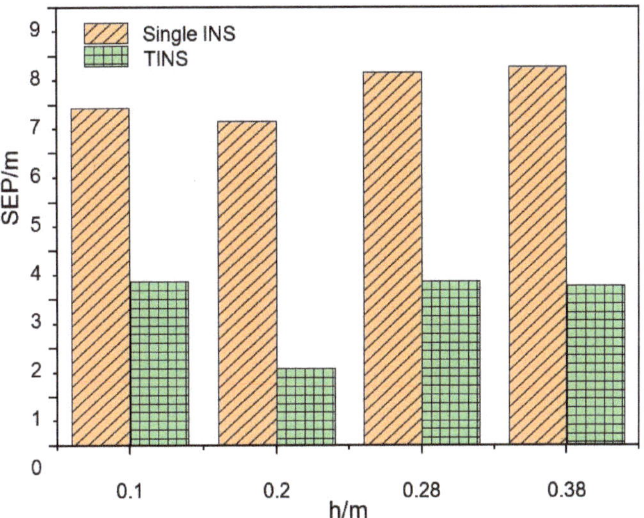

Figure 9. Comparison of positioning accuracy in the inter-INS distances influence experiments.

4.2. The Tri-INS Plane Spatial Position Influence Experiments

In the tri-INS plane spatial position influence experiments, the SEP of single-INS positioning in the positive plane is 7.77 m, 6.57 m, 6.76 m, 7.64 m and 6.67 m, respectively, while the SEP of TINS positioning is 2.38 m, 2.52 m, 2.24 m, 3.73 m and 2.91 m, respectively (as shown in Figure 10a). The positioning accuracy of the TINS is improved by 69.36%, 61.64%, 66.86%, 51.17% and 56.37% compared to the single-INS. In the positive plane, the positioning accuracy of TINS is highest when α_3 is 90°, and the positioning accuracy improvement effect is best when α_3 is 0°. In the negative plane, the SEP of single-INS positioning is 6.19 m, 6.76 m, 6.16 m, 6.48 m and 5.93 m, respectively, while the SEP of TINS positioning is 2.49 m, 1.72 m, 3.00 m, 2.20 m and 2.22 m, respectively (as shown in Figure 10b). The positioning accuracy of TINS is improved by 59.77%, 74.56%, 51.30%, 66.05% and 62.56% compared to the single-INS. The positioning accuracy is the highest and the positioning accuracy improvement effect is also the best when α_3 is 45°.

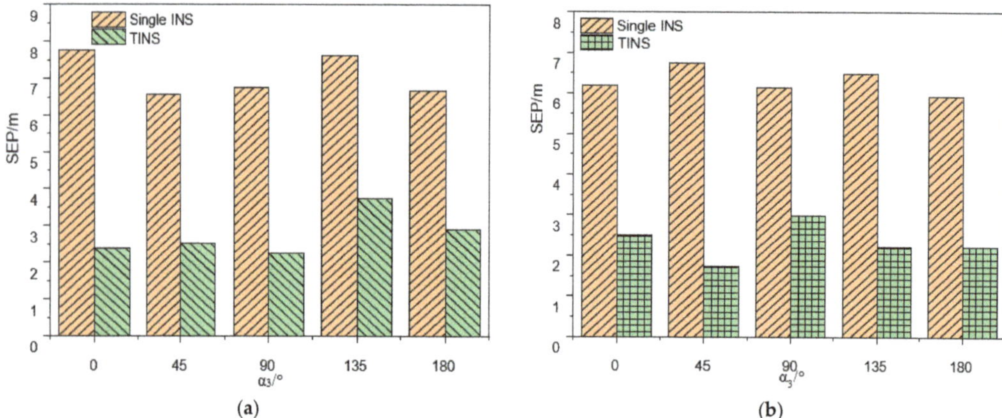

Figure 10. Comparison of positioning accuracy in the tri-INS plane spatial position influence experiments: (**a**) Positive plane; (**b**) Negative plane.

The experimental results shown in Figures 9 and 10 as well as the analysis results of the previous simulations carried out by the authors of [10] show that the positioning accuracy of the TINS is higher than that of the single-INS under various working conditions. The real triangle formed by the three INS can be approximated as a rigid body. The lengths of the three sides of the triangle are sent to EKF as the observed quantities, which plays a role in limiting the accumulation of INS errors. When the inter-INS distances are 0.2 m and α_3 is 45° in the negative plane, the positioning accuracy is the highest and the positioning accuracy improvement effect is also the best. However, considering the positioning accuracy and the convenience of the equipment installation, the positive plane with α_3 90° (horizontal installation) is optimum.

The estimated outputs of the three INS positionings were sampled every 20 s, using the EKF in Experiment No.7 of Table 2, and were plotted as a triangle (estimated triangle) in three-dimensional space at each sampling. It was projected on a north–east plane and east–up plane as shown in Figure 11. Time 0 was the initial time at which the projections in the two planes completely coincided with the real position projections of the three INS. In the later times, the shape of projections in the two planes changed to some extent with the movement of the experimental device. Figure 12 is the real-time variation diagram of the estimated triangle side length under this experimental condition. It can be seen from the figure that the estimated triangle side lengths have changed to some extent, but it is not as obvious as the change in projections in the two planes. It indicates that the shape of the triangle in three-dimensional space has not changed that much. The main reason for the change in the shape of the projections in the two planes is that the normal vector of the tri-INS plane deviated to a large extent. The secondary reason is the change in shape of the estimated triangle. It can be seen from the EKF model of the TINS positioning of the shearer that the velocity of the shearer is collected by the coder, so the main factors affecting the positioning accuracy are the accuracy of the three INS positioning estimated outputs and the accuracy of the observed values at the last moment. The deviation of the plane normal vector of the estimated triangle and its shape change mainly depend on the output of the estimated values of the three INS at the last moment from the EKF. When the three INS are installed, the real inter-INS distances will not change, since the accuracy of the observed values has been determined. Therefore, the main factor influencing the positioning accuracy is the accuracy of the estimated values of the three INS positions at the last moment.

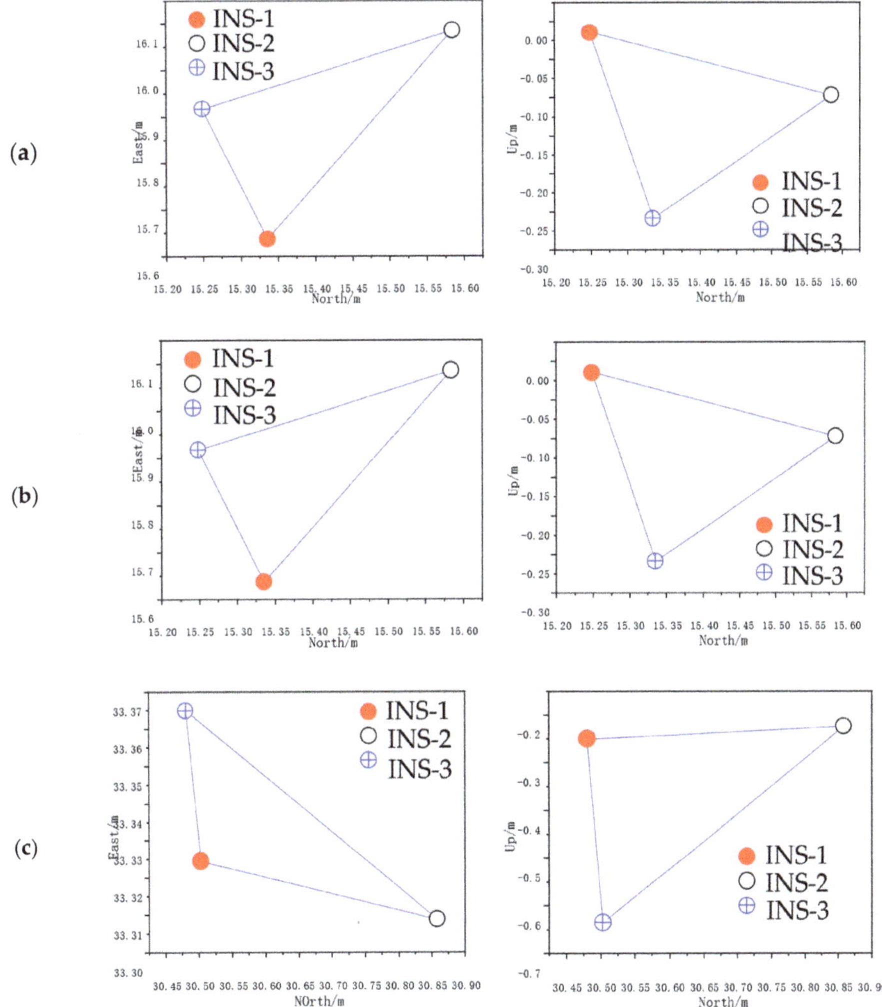

Figure 11. Projections of estimated triangle on east-north plane (left) and north-up plane (right): (**a**) 0 s; (**b**) 20 s; (**c**) 40 s.

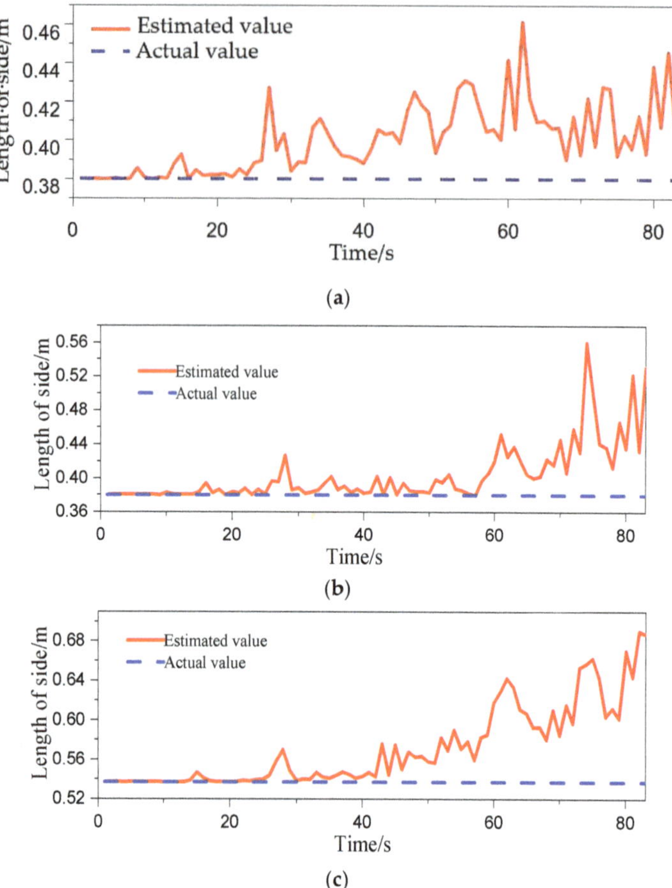

Figure 12. Variation of the side length of the estimated triangle. (**a**) r_{12}; (**b**) r_{13}; (**c**) r_{23}.

5. Conclusions

In this paper, an experimental device was built according to the mathematical model of TINS positioning of shearers, and the influence of inter-INS distances and the spatial position of the tri-INS plane was analyzed experimentally. The results show that the positioning accuracy of TINS is higher than that of single-INS and the positioning trajectories are smoother under various experimental conditions. The main reason affecting the positioning accuracy is the output accuracy of the three INS position estimated values from the EKF at the last moment. Considering positioning accuracy of the shearer and the convenience of the equipment installation, the installation parameters of a positive plane with α_3 90° (horizontal installation) and 0.2 m inter-INS distances are the most suitable. Although the experimental device could not be operated for a long time due to the limitations of the overground experimental conditions, the experiments in this paper and the author's previous simulations [10] show that the positioning accuracy of multi-INS is higher than that of single-INS with the same type. In addition, the purpose of this paper is to verify the effectiveness of the proposed method for improving the positioning accuracy of shearers. If the accuracy improvement effect is verified, high-precision INS will be used in real shearer experiments to meet the application requirements. It is also necessary to consider the influence of the vibration of running the shearer on the positioning accuracy during underground experiments in mines. Whether vibration isolation technology should be used to eliminate or reduce these influences needs further study. Multi-INS positioning technology not only plays a positive role in improving positioning accuracy, but the system also has the advantages of strong fault tolerance and good reliability. This research on multi-INS positioning technology of shearer has a good prospect of application in the process of intelligent coal mining.

Author Contributions: Conceptualization, S.W.; methodology, S.W. and C.L.; validation S.W. and C.L.; software, K.S., W.D. and W.L; writing—original draft preparation, C.L.; review and editing, S.W., K.S., W.D. and W.L.; funding acquisition, S.W. and W.L. All authors have read and agreed to the published version of the manuscript.

Funding: This research was funded by Funds of the Natural Science Foundation of China, grant number 51874279, China Scholarship Council, grant number 202208260074 and Education Key Project of Anhui Provincial Department, grant number 2022AH051636.

Data Availability Statement: The data presented in this study are available on request from the corresponding author.

Conflicts of Interest: The authors declare no conflict of interests.

References

1. Ge, S.; Hao, X.; Tian, K. Principle and key technology of autonomus navigation cutting for deep coal seam. *J. China Coal Soc.* **2021**, *46*, 774–788.
2. Ge, S.; Wang, S.; Cao, B. Autonomous positiong principle and technology of intelligent shearer and conveyor. *J. China Coal Soc.* **2022**, *47*, 75–86.
3. Wang, S.; Ge, S.; Wang, S. Development and chanllege of unmanned autonomous longwall fully-mechanized coal ming face. *Coal Sci. Technol.* **2022**, *50*, 231–243.
4. Reid, D.C.; Dunn, M.T.; Reid, P.B. A practical inertial navigation solution for continuous miner automation. *Coal Oper. Conf.* **2012**, *12*, 114–119.
5. Dunn, M.T.; Thompson, J.P.; Reid, P.B. High accuracy inertial navigation for underground mining machinery. In Proceedings of the IEEE International Conference on Automation Science and Engineering (CASE), Seoul, Republic of Korea, 20–24 August 2012; pp. 1179–1183.
6. Ralston, J.C.; Reid, D.C.; Dunn, M.T. Longwall automation: Delivering enabling technology to achieve safer and more productive underground mining. *Int. J. Min. Sci. Technol.* **2015**, *25*, 865–876. [CrossRef]
7. Li, A.; Hao, S.; Wang, S. Experimental Study on Shearer Positioning Method Based on SINS and Encoder. *Coal Sci. Technol.* **2016**, *44*, 95–100.
8. Wang, S.; Wang, S.; Boyuan, Z. Dynamic Zero-velocity Update Technology to Shearer Inertial Navigation Positioning. *J. China Coal Soc.* **2018**, *43*, 578–583.
9. Wang, S.; Zhang, B.; Wang, S. Dynamic Precise Positioning Method of Shearer Based on Closing Path Optimal Estimation Model. *IEEE Trans. Autom. Sci. Eng.* **2018**, *16*, 1468–1475.
10. Cheng, L.; Wang, S.; Ge, S. Redundant Multi-INS positioning algorithm of shearer and analysis of its rationality. *J. China Coal Soc.* **2019**, *44*, 746–753.
11. Wang, S. Research on shearer positioning with double-INS. *Sens. Rev.* **2019**, *39*, 577–584. [CrossRef]
12. Wang, S. Research on Shearer Positioning Technology with Heterogeneous Multi-Source Information Fusion. Ph.D. Thesis, China University of Minging and Technology, Xuzhou, China, 2022.
13. Cao, B. Research on Fusion Localization Approach of IMU and Ultra-Wideband for Shearer. Ph.D. Thesis, China University of Minging and Technology, Xuzhou, China, 2022.
14. Cao, B.; Wang, S.; Ge, S. Research on Sheaer Positioning Experiment Based on IMU and UWB at the end of Underground Coal Mining Working Face. *Coal Sci. Technol.* **2022**, *51*, 217–228. [CrossRef]
15. Wang, S.; Wang, S. Improving the shearer positioning accuracy using the shearer motion constraints in longwall panels. *IEEE Access* **2020**, *8*, 52466–52474. [CrossRef]
16. Reid, D.; Ralston, J.; Dunn, M. Longwall shearer automation: From research to reality. In *Machine Vision and Mechatronics in Practice*; Springer: Berlin/Heidelberg, Germany, 2015; pp. 49–57.
17. Ralston, J.; Reid, D.C.; Hargrave, C. Sensing for advancing mining automation capability: A review of underground automation technology development. *Int. J. Min. Sci. Technol.* **2014**, *24*, 305–310. [CrossRef]
18. Liu, Z.; Liu, H.; Wu, X. The Research on Dynamic Calibration for Accelermeter Zero of Shipborne Double INS. *Radio Eng.* **2015**, *45*, 30–33.
19. Zhang, L.; Wang, X. Research on data dispose of multiple INS configuration navigation system on ships. *Ship Sci. Technol.* **2021**, *43*, 176–179.

20. Bai, J.; Chang, M.; Wang, H. Simulation Study and Analysis on the Multi-Inertial Navigation Fusion Scheme in Aircraft Navigation System. *Unmanned Syst. Technol.* **2020**, *3*, 79–91.
21. Si, F.; Zhao, Y.; Zhang, X. The estimation of wing flexure deformation in transfer alignment based on inertial sensors network. In Proceedings of the IEEE Aerospace Conference, Big Sky, MT, USA, 4–11 March 2017; pp. 1–11.

Article

A Self-Oscillating Driving Circuit for Low-Q MEMS Vibratory Gyroscopes

Tian Han [1], Guanshi Wang [2], Changchun Dong [1], Xiaolin Jiang [1], Mingyuan Ren [1] and Zhu Zhang [1,*]

[1] Department of Artificial Intelligence, Jinhua Advanced Research Institute, Jinhua 321013, China; htopen@foxmail.com (T.H.); hitdongcc@163.com (C.D.); jlynner@163.com (X.J.); rmy2000@126.com (M.R.)
[2] Development Strategy Research Center, China Electronics Technology Group, Beijing 100846, China; wanggs@163.com
* Correspondence: zhuzhang.zz@foxmail.com; Tel.: +13796604991 or +86-0579-82295992

Abstract: This article establishes a circuit model with which to analyze the difficulty of auto-gain control driving for low-Q micromechanical gyroscopes at room temperature and normal pressure. It also proposes a driving circuit based on frequency modulation to eliminate the same-frequency coupling between the drive signal and displacement signal using a second harmonic demodulation circuit. The results of the simulation indicate that a closed-loop driving circuit system based on the frequency modulation principle can be established within 200 ms with a stable average frequency of 4504 Hz and a frequency deviation of 1 Hz. After the system was stabilized, the root mean square of the simulation data was taken, and the frequency jitter was 0.0221 Hz.

Keywords: MEMS; vibratory gyroscopes; self-oscillating driving circuit; automatic gain control; frequency modulation drive circuit

Citation: Han, T.; Wang, G.; Dong, C.; Jiang, X.; Ren, M.; Zhang, Z. A Self-Oscillating Driving Circuit for Low-Q MEMS Vibratory Gyroscopes. *Micromachines* **2023**, *14*, 1057. https://doi.org/10.3390/mi14051057

Academic Editor: Huiliang Cao

Received: 20 April 2023
Revised: 11 May 2023
Accepted: 15 May 2023
Published: 16 May 2023

1. Introduction

Over the last decade, micro-electromechanical system (MEMS)-based inertial sensors have undergone widespread developments in both research efforts and commercial products, offering advantages such as low cost, small size, low power consumption and suitability for batch fabrication [1–4].

In general, the gyroscopes are driven by AC voltage. This approach often involves utilizing an external oscillator to generate a driving signal that aligns with the resonant frequency of the mechanical gyroscope driving direction. To achieve frequency matching, the oscillator's frequency is manually adjusted [5]. However, this method is often inaccurate, particularly when the gyroscope's parameters fluctuate due to changes in the external environment. To tackle this issue, a closed-loop driving method has been introduced to facilitate automatic frequency matching between the driving circuit and the mechanical structure [6,7]. The benefit of this implementation method is that it averts frequency imbalances caused by structural parameter shifts resulting from variations in the external environment. As the quality factor of the micromechanical gyroscope structure is significantly high, it operates akin to embedding a narrowband band-pass filter function in the system [8]. The closed-loop drive system capitalizes on the frequency selection capability of the structure to achieve automatic tracking and matching of frequencies [9]. This filter significantly attenuates signals whose frequencies are dissimilar to the resonant frequencies of the structure drivers while amplifying signals whose frequencies are equivalent to this resonant frequency, culminating in the formation of a stable closed-loop driving signal [10].

In the gyroscope driving circuit, both the loop gain and the loop phase are related to the quality factor Q [11]. The Q value of the micromechanical gyroscope in the atmospheric environment is only a few hundred [12]. In this case, the loop gain of the circuit is large and other frequency signals in the environmental noise are also amplified, resulting in the frequency of the output signal of the gyroscope closed-loop drive circuit and the driving mode [13]. There is a deviation in the natural frequency of the state, and the frequency jitter is also large [14]. To resolve this problem, the micromachined gyro sensor structure is usually vacuum packaged to improve its Q value; however, this increases the manufacturing costs [15]. The traditional gyro closed-loop drive circuit is generally based on the principle of automatic gain control [16]. When driven by this principle, when the micromechanical gyro is in the working state, its structure is in a resonant state, and the driving voltage signal will be coupled with the driving direction detection terminal via parasitic capacitance [17]. It causes the amplitude and phase of the driving direction detection signal to change so that the self-excited driving conditions of the system cannot be satisfied [18].

This article presents a driving circuit that utilizes the frequency modulation principle to eradicate the identical frequency coupling from the driving signal to the displacement signal. Moreover, this circuit is harmonious with low-Q-value gyroscope structures.

This paper is organized as follows. Section 2 illustrates the MEMS gyroscope structure and principle. Section 3 illustrates the automatic gain control circuit. Section 4 describes the frequency modulation drive circuit and implementation details. The results of the simulation are presented and discussed in Section 5, and the conclusions are listed in Section 6.

2. MEMS Gyroscope Structure and Principle

MEMS vibratory gyroscopes are generally composed of a proof mass with elastic beams and corresponding combs in both directions for capacitance detection. Figure 1 shows a standard model of the sensitive structure for MEMS gyroscopes with a single proof mass driven in the x-direction [19].

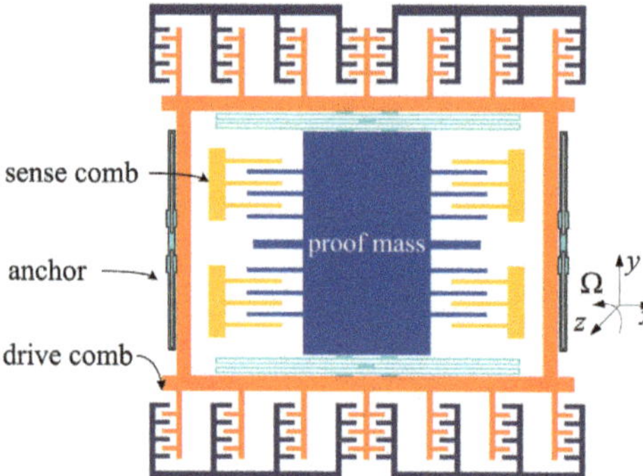

Figure 1. MEMS vibratory gyroscope structure.

The gyroscope operates in two distinct modes: driving and detecting. In the driving mode, a stable driving signal is added to the gyro structure, producing Coriolis force from the angular velocity signal in the z-axis direction. This displacement generates changes in the detection capacitance along the y-axis. In the detecting mode, the detected capacitance displacement (or angular velocity signal) is converted into a voltage signal output via the interface circuit [20].

2.1. Drive Mode

The closed-loop drive mode is commonly utilized in micromechanical gyroscopes. After undergoing processes such as peak detection and variable gain, the driving electrode's vibration signal is fed back to the driving electrode to create a closed-loop drive signal [21].

Figure 2 depicts the sensitive combs' driving schematic. As the proof mass moves, the capacitance between the fixed combs and the proof mass changes accordingly. If a DC voltage is applied between the electrodes of the fixed combs and the proof mass, the current can be calculated as

$$i = \frac{dQ}{dt} = \frac{d(CV)}{dt} = C\frac{dV}{dt} + V\frac{dC}{dt} \tag{1}$$

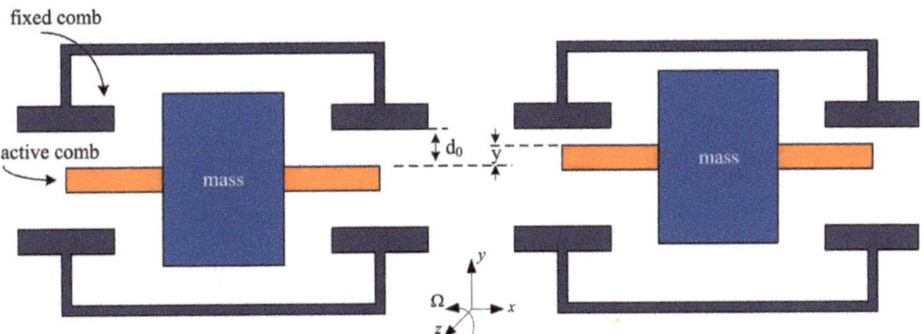

Figure 2. Structure of drive-sensing combs.

Figure 2 shows the DC voltage between the fixed comb teeth and the electrodes on the mass block. If the first term to the right of Equation (1) is zero, the number of mass blocks is

$$i = N_2 V_p \frac{dC}{dt} = N_2 V_p \frac{dC}{dx}\frac{dx}{dt} = N_2 V_p \varepsilon \frac{z}{y}\frac{dx}{dt} \tag{2}$$

Equation (2) demonstrates that the output current of the micromechanical gyroscope's driving sensitive comb tooth in the x-axis direction is directly proportionate to the speed.

2.2. Sense Mode

The vibration of the micromechanical gyroscope on the x-axis results in the generation of Coriolis force acceleration on the y-axis via the angular velocity in the z-axis. When the sense combs vibrate on the y-axis, the equivalent capacitance changes [21]. The principle underlying the detection of angular velocity is based on the detection of alterations in the equivalent capacitance on the y-axis. The amplitude of capacitance vibration is directly proportional to the input angular velocity.

The proof mass experiences no Coriolis force and the sense comb remains central when the external input angular velocity is zero. In addition, the distances between the fixed combs on the upper and lower sides are equal, and thus the values of C1 and C2 are also equal. At this point, the sensor is in the zero position, as depicted in the left portion of Figure 3. However, when the external angular velocity signal is not zero, the proof mass experiences Coriolis force, causing the mass block to shift along the detection axis. This movement leads to variations in the capacitance values on both sides, as illustrated on the right side of Figure 3.

Figure 3. Structure of sense combs.

The dynamic equation of sense direction is

$$M_s \frac{d^2 y}{dt^2} + \lambda_s \frac{dy}{dt} + K_s y = 2B_d M_s \Omega_0 \omega \cos \omega_i t \cos(\omega t + \varphi_d) \tag{3}$$

where the proof mass is represented by M_s and λ_s is the damping force coefficient acting in the detection axis direction. K_s is the elastic coefficient in the same direction; y is the vibration displacement detected in the detection axis direction.

To obtain the angular velocity signal with the maximum amplitude, we selected $\cos(\omega t + \frac{\varphi_{s1} + \varphi_{s2}}{2} + \varphi_d)$ as the reference signal for demodulation, and the output after demodulation was

$$y_1(t) = \tfrac{1}{2}(B_{s1} + B_{s2}) \cos(\omega_i t + \tfrac{\varphi_{s2} - \varphi_{s1}}{2})\{1 + \cos[2(\omega t + \tfrac{\varphi_{s1} + \varphi_{s2}}{2} + \varphi_d)]\}$$
$$+ \tfrac{1}{2}(B_{s1} - B_{s2}) \sin(\omega_i t + \tfrac{\varphi_{s2} - \varphi_{s1}}{2}) \sin[2(\omega t + \tfrac{\varphi_{s1} + \varphi_{s2}}{2} + \varphi_d)] \tag{4}$$

The frequency of the driving signal for the micromechanical gyroscope is typically several hundred times greater than the input angular velocity frequency. This discrepancy allows for the simplification of the signal, represented in Formula (4). A low-pass filter can effectively remove the higher-frequency component present in the signal expression.

$$y_2(t) = \frac{1}{2}(B_{s1} + B_{s2}) \cos(\omega_i t + \frac{\varphi_{s2} - \varphi_{s1}}{2}) \tag{5}$$

The frequency of the output signal is equal to that of the angular velocity signal. Furthermore, there is direct proportionality between the amplitude of the output signal and the angular velocity signal. Such a correlation enables the measurement of the angular velocity signal by detecting the corresponding amplitude and frequency of the output signal.

2.3. Closed-Loop Driving Stability Analysis

The block diagram of the closed-loop self-excited drive system is depicted in Figure 4. A positive feedback loop is formed by the I/V converter, adder, F/V converter and gyroscope structure, resulting in the self-excited oscillation of the gyro. Meanwhile, a feedback control loop is established by the peak detection circuit and the proportional integral (PI) controller to ensure a stable driving amplitude. The PI controller regulates the amplitude of vibration while the F/V converter stabilizes the amplitude of the Vac. The gyroscope in the X-axis direction is

$$M_d\left(\frac{d^2 x}{dt^2} + 2\zeta_d \omega_d \frac{dx}{dt} + \omega_d^2 x\right) = K V_{dc} \frac{dx}{dt} \tag{6}$$

where M_d is the equivalent mass in the drive axis, ω_d is the natural frequency of the drive mode, ζ_d is the damping ratio and K is the loop gain.

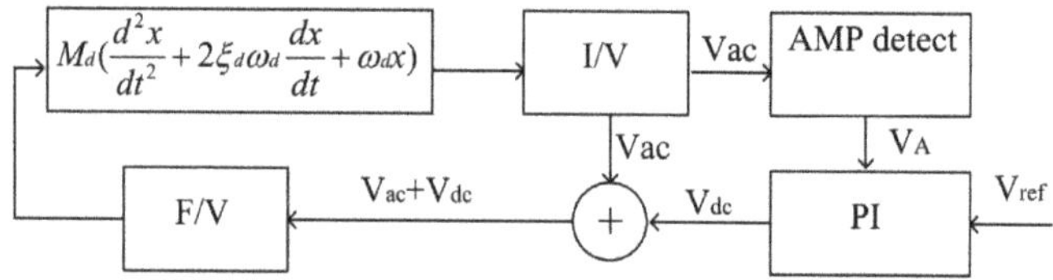

Figure 4. Closed-loop self-excited drive system.

The peak detection circuit comprises an amplitude detection circuit and a low-pass filter. The electrical differential equation of the peak detection circuit is represented by Equation (3), where K_1 denotes the coefficient of the amplitude detection circuit, V_A

represents the output signal of the peak detection circuit and τ_1 is the time constant of the low-pass filter.

$$|K_1 V_{ac}| = V_A + \tau_1 \frac{dV_A}{dx} \tag{7}$$

The PI controller is composed of a proportional amplification component and an integral component. The electrical differential equation for the PI controller is represented by Equation (8). Notably, K_2 serves as both the coefficient for the proportional part and the integral time constant.

$$-\frac{dV_{dc}}{dx} = K_2 \frac{dV_A}{dx} + \tau_2(V_A - V_{ref}) \tag{8}$$

Equations (6)–(8) form a system of differential equations, which describes a nonlinear system. Solving the differential equations allows for the determination of the initial vibration condition of the system. The solution system's characteristic root is

$$S = -\frac{1}{2\tau_1}\left(1 \pm \sqrt{1 - 2\tau_1\left(\frac{KV_{ref}}{M_d} - \xi_d\omega_d\right)}\right) \tag{9}$$

So, the initial vibration condition of the system is

$$\frac{KV_{ref}}{M_d} - \xi_d\omega_d > 0 \tag{10}$$

$$\tau_1 \tau_2 < K_2 \tag{11}$$

The driving force of the gyroscope, as defined by Formula (11), must exceed the damping force. The stability of a nonlinear system can be determined by analyzing the real part of its characteristic roots; a negative real part indicates stability with a consistent post-vibration amplitude. When the conditions required to initiate vibration are met, a stable vibration amplitude can be achieved.

When $2\tau_1\left(\frac{KV_{ref}}{M_d} - \xi_d\omega_d\right) > 1$, overshoot and oscillation will be generated, and increasing τ_1 will increase the duration of the overshoot and oscillation phenomenon. The oscillation degree will be deepened, resulting in the instability of the system. When $2\tau_1\left(\frac{KV_{ref}}{M_d} - \xi_d\omega_d\right) < 1$, the amplitude of the gyro-driven signal will converge to the stable value at the speed of $2\tau_1$. Increasing KV_{ref} can shorten the starting time.

3. Analysis of Automatic Gain Control Circuit

The principle is shown in Figure 5. Capacitor C_2 is the parasitic capacitance between the driving end of the gyro structure and the driving detection end. The driving voltage V_{drive} is converted into current i_2 through the capacitor C_2. The frequency of this current is the same as the frequency of the driving voltage. The phase difference in the current is 90°, and the expressions of i_1 and i_2 can be expressed as follows:

$$i_1 = A\sin\omega t \tag{12}$$

$$i_2 = B\cos\omega t \tag{13}$$

The feedback current i_3 of the charge amplifier is as follows:

$$i_3 = i_1 + i_2 = A\sin\omega t + B\cos\omega t = \frac{1}{\sqrt{A^2 + B^2}}\sin(\omega t + \theta) \tag{14}$$

$$\theta = \operatorname{arctg}\frac{B}{A} \tag{15}$$

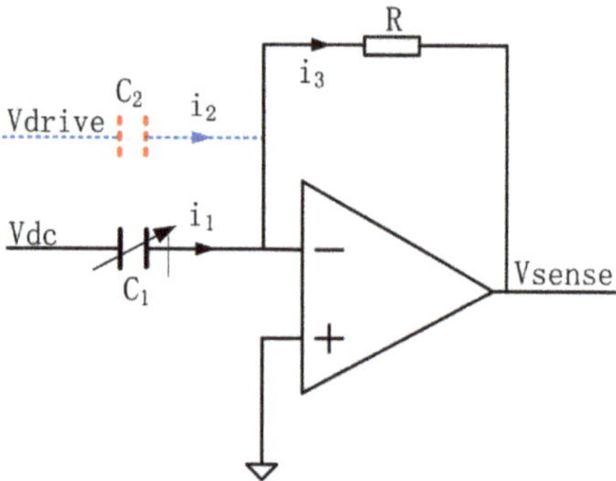

Figure 5. Schematic diagram of driving signal coupling.

The parasitic capacitance C_2 causes the phase and amplitude of the feedback current of the charge amplifier at the driving end to change. The phase difference between the feedback current and i_1 changes the angle, and the angle value is determined by the ratio of the parasitic capacitance to the differential capacitance. According to the classic oscillation principle, the closed-loop self-excited drive requires that the phase difference between the drive signal and the charge amplifier meet certain conditions and must be smaller than a predetermined low value [22,23]. Exceeding this value will cause the oscillation phase condition to be inconsistent and the circuit will not oscillate. Therefore, to reduce the impact of parasitic capacitance C_2, the automatic gain control circuit needs to decrease the loop gain or choose a high Q value gyroscope structure [24,25].

From the perspective of sensor head design, good packaging, preferably vacuum packaging, is required. Therefore, the head of the normal temperature and pressure package should be designed using a frequency modulation scheme. After adopting this driving scheme, the frequency of the driving voltage changes, differing from the frequency of the sensitive current after square wave modulation. In this way, the frequency separation between the AC driving voltage signal and the sensitive current signal can be realized, the problem that the closed-loop system cannot self-oscillate can be solved and the environmental interference caused by the large loop gain can also be reduced.

4. Frequency Modulation Drive Circuit

The principle of the frequency modulation drive circuit is shown in Figure 6. The sensitive current is converted into a voltage signal through the charge amplifier. The phase difference between the voltage signal and the sensitive current signal is $90°$. The phase shift of the pre-stage charge amplifier is offset by the post-stage $90°$ phase shifter to meet the phase condition of the closed-loop drive. The signal is amplified by a first-stage amplifier circuit to output three signals. All the way through the peak detection circuit, PI controls the DC voltage output. After the other two paths pass through the follower and the inverter, the AC signals are obtained. The three-way signals pass through two adders to obtain signals, are multiplied by the high-frequency carrier and then they output the drive signals of the micromechanical gyroscope to achieve closed-loop driving.

4.1. Charge Amplifier

Figure 7 is a circuit structure diagram of a charge amplifier based on a folded cascode three-stage operational amplifier. The equivalent input noise density of this amplifier is as follows:

$$V_n^2 = \frac{16kT}{3}\left(\frac{1}{g_{m2}} + \frac{g_{m5} + g_{m8}}{g_{m2}^2}\right) \tag{16}$$

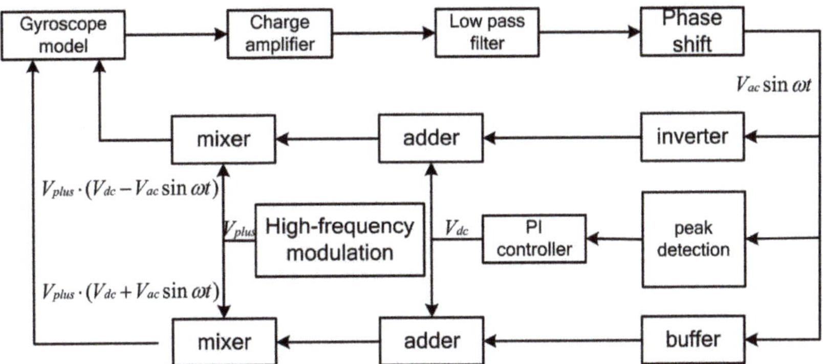

Figure 6. Frequency modulation drive circuit.

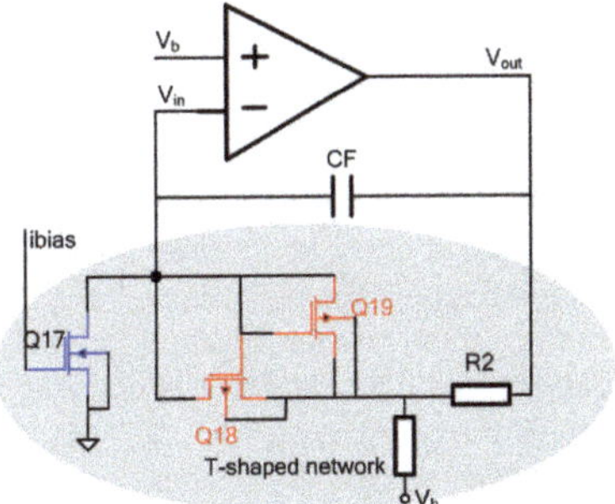

Figure 7. Charge amplifier.

The charge amplifier uses a T-shaped network structure to increase the transimpedance gain, as shown in Figure 4. Transistors Q17~Q19 and resistors R1 and R2 form a T-shaped resistor network larger than 107 Ω. Using Q17 and Q19 allows Q18 to achieve a smaller VGS and reduces the W/L of Q18; so, Q18 is in the linear region. Its equivalent resistance exceeds 1 MΩ and is proportional to the bias resistance R in the bias circuit, without significant changes in time and temperature. Capacitor CF is about 0.3 pF, which can improve the stability of the system. The equivalent resistance is shown in Formula (6):

$$R_{eq} = R_M\left(1 + \frac{R_2}{R_1}\right) + R_2 \tag{17}$$

R_M is the equivalent resistance of Q18. If R_M is greater than 1 MΩ, it will become the main noise source of the charge amplifier. The charge amplifier output signal-to-noise ratio SNR determined by the equivalent resistance of the transistor can be expressed as follows:

$$SNR = \sqrt{\frac{I_{IN}^2 R_{eq}}{4KT(R_2/R_1 + 1)}} \approx \sqrt{\frac{I_{IN}^2 R_M}{4KT}} \tag{18}$$

where I_{IN} is the charge amplifier input current:

$$I_{IN} = \frac{\partial C}{\partial t} V_b = C_0 V_b \omega \sin \omega t \tag{19}$$

C_0 is the maximum variation in the sensitive capacitance at the drive end, ω is the resonant frequency and V_b is the forward bias voltage.

4.2. Phase Shifter Circuit

The phase shifter circuit consists of an adder and two integrators, as shown in Figure 8. It can carry out the 90° phase shift of the input signal and realize the conversion of DC bias voltage. The resistance values of the three resistors of the adder are equal, and the integrator is divided into a feedback integrator and a feedforward integrator. After the latter achieves a 90° phase shift, the output current value is fed back to the adder via the feedback integrator.

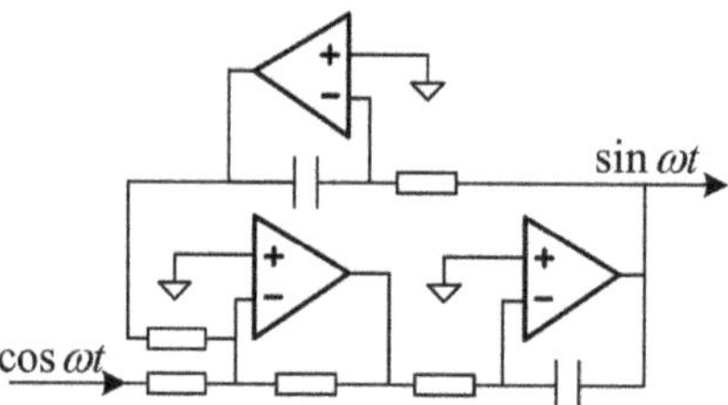

Figure 8. Phase shifter circuit.

4.3. PI Controller Circuit

The PI controller circuit adopts the structure shown in Figure 9, in which the operational amplifier and capacitor C6 form an integrator, the output of the peak detection circuit after passing through the filter is connected to the negative input terminal of the operational amplifier through the resistor R9 and its positive input terminal is connected to the DC comparison voltage V_{ref}. The PI controller adjusts the DC part of the driving signal. The working mode can be analyzed using a DC comparator. Due to the "virtual short" of the operational amplifier, the voltage of node A is equal to V_{ref}. If the voltage difference between node V_{out} and node A is the same as when it is zero, there is a constant current flowing through the resistor R9, and the output node voltage changes after the current is integrated by C6. When the integrated current value flowing through C6 is zero, the system enters a steady state and the output node voltage is equal to the voltage of node A.

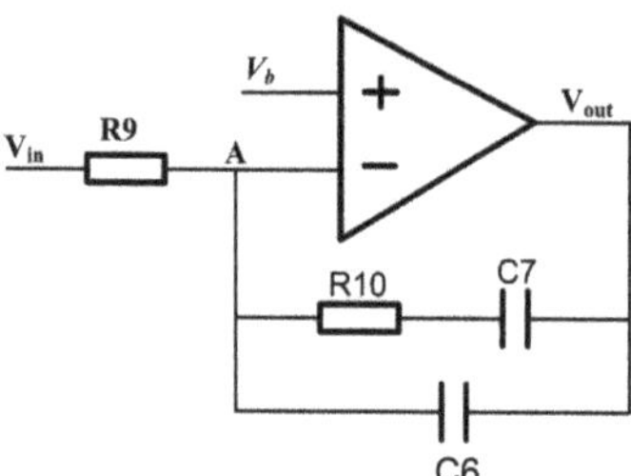

Figure 9. PI controller.

4.4. Mixer Circuit

To avoid the influence of the offset voltage of the operational amplifier in the mixer, the temperature coefficient of the offset voltage should be reduced. The topology of the three-stage operational amplifier is adopted, the second-stage capacitor is compensated and the switch unit adopts a 6 MOS structure, which effectively eliminates clock feedthrough and charge injection, as well as lower on-resistance.

As shown in Figure 10, one end of the switch S1 is connected to the positive end of the operational amplifier, while the other end is connected to the reference voltage V_{ref}, and is turned on and off by a square wave signal. The signal at the positive input of the mixer can be expressed as follows:

$$U_2(t) = \begin{cases} V_{ref} & nT_C < t < \left(n + \frac{1}{2}\right)T_C \\ Us(t) & \left(n + \frac{1}{2}\right)T_C < t < (n+1)T_C \end{cases} \tag{20}$$

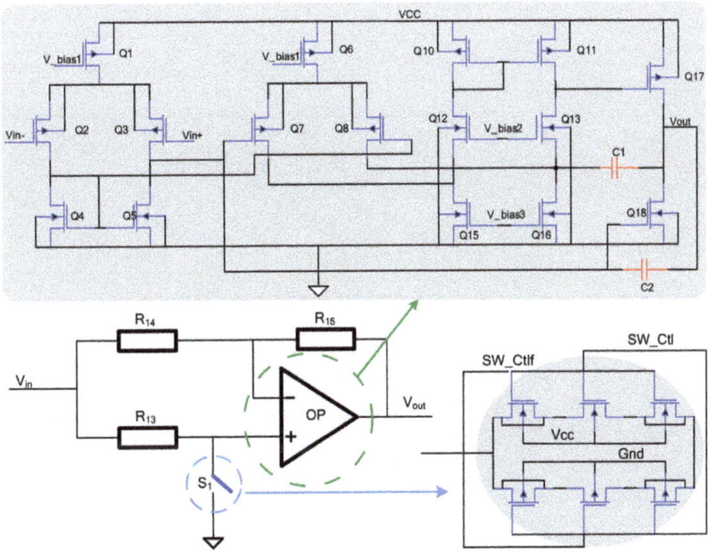

Figure 10. Mixer circuit.

If the resistors R14 and R15 are equal, then the output waveform of the mixer is as follows:

$$U_o(t) = -U_s(t) + 2U_2(t) \tag{21}$$

Substituting Formula (9) into (10), Expression (11) can be obtained:

$$U_o(t) = \begin{cases} 2V_{ref} - U_s(t) & nT_C < t < \left(n + \frac{1}{2}\right)T_C \\ U_s(t) & \left(n + \frac{1}{2}\right)T_C < t < (n+1)T_C \end{cases} \tag{22}$$

5. Results

Table 1 shows the structural parameters of the silicon gyroscope employed in this paper.

Table 1. Structural parameters of the silicon gyroscope.

Parameter		Unit
Drive mass M_d	2.1493	$Kg \times 10^{-7}$
Damping coefficient	2.1247	$\lambda_d \times 10^{-5}$
Elastic coefficient	1.7219	$Kd \times 10^2$
Central capacitance	0.9445	pF
Resonant frequency	4.5047	KHz
Quality factor	205.7238	

The monolithic integrated design of the MEMS interface circuit is based on 0.5 μm 18 V CMOS process technology. To reduce the low-frequency 1/f noise and thermal noise of the charge amplifier, the charge amplifier input transistor uses a large aspect ratio. To prevent the AC signal of the driving circuit affecting the sense circuit, the driving and the sense circuit module become separate modules. Meanwhile, a grounded metal wire is added between the driving circuit and the sense circuit, reducing the interference of the driving signal in the sensitive circuit. In addition, the charge amplifier input tube used in the interface ASIC provides a fully symmetrical layout, which can reduce the amplifier's offset voltage. The interface circuit chip area is about 5.05 mm × 3.7 mm, and the overall interface circuit includes the gyroscope drive circuit, the sense circuit and the quadrature circuit. The interface ASIC contains 818 transistors, 54 resistors, 60 capacitors and 5 diodes. The layout of the MEMS gyroscope interface ASIC is shown in Figure 11. The gyroscope interface ASIC operates at ±9 V supply voltage and its power consumption is 360 mW.

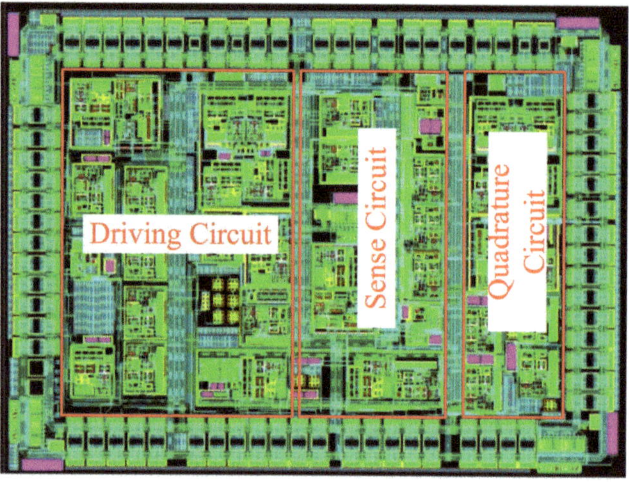

Figure 11. Layout of closed-loop micromachined gyroscope interface ASIC.

5.1. System-Level Modeling and Simulation

The performance of the driving circuit is also one of the key factors limiting the performance of the gyroscope system. The MEMS gyroscope interface circuit adopts the automatic gain control module to achieve stable amplitude control of the driving voltage, ensuring the self-excited oscillation of the gyroscope under normal temperature and a low Q value. The driving circuit allows for complete closed-loop self-excited oscillation, and the sense circuit can detect the input angular velocity information. After the driving circuit stabilizes the self-excited amplitude, the sense circuit can complete the demodulation and output of the angular velocity signal. Therefore, it is crucial for the MEMS gyroscope's driving circuit to achieve self-excited oscillation. In this work, a system-level simulation model of the gyroscope driving loop was established using SIMULINK, as shown in Figure 12.

The automatic gain module ensures that the driving circuit has stable amplitude and is quick to start, allowing the gyroscope system to quickly enter into operation. By changing the value of the set reference voltage, the oscillation amplitude of the gyroscope drive circuit can be altered. Figure 13 shows the simulation diagram of the self-excited start of the driving circuit of the MEMS gyroscope. The simulation results show that the driving circuit completes the self-excitation oscillation in about 0.2 s, and the resonant frequency of the driving voltage signal is 4504 Hz. Compared with the non-self-excited oscillation method, the system does not need to readjust the locking frequency and phase when the resonant frequency of the gyroscope changes. Therefore, the closed-loop self-excited drive scheme based on the secondary demodulation principle is more adaptable and consumes less hardware resources.

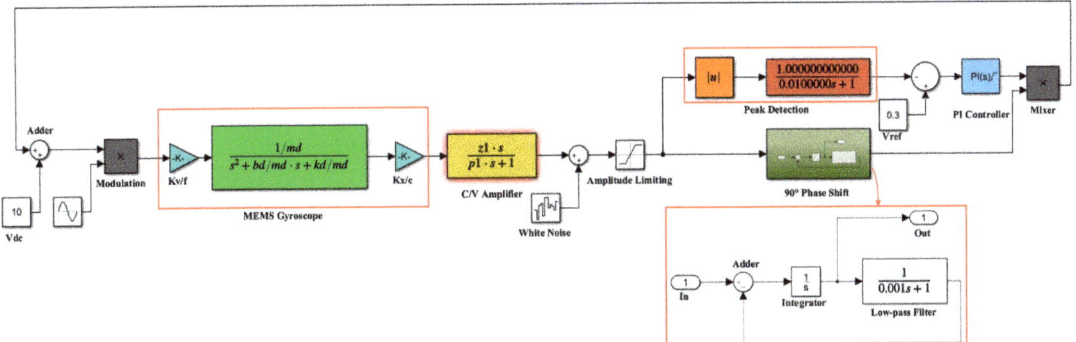

Figure 12. Driving circuit system-level simulation model diagram.

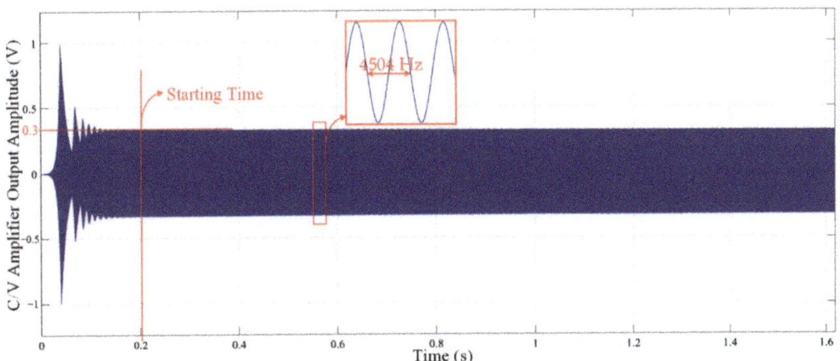

Figure 13. The simulation diagram of the self-excited start of the driving circuit of the MEMS gyroscope.

5.2. Transistor-Level Simulation Analysis

Figure 14 is the overall simulation result of the mixer. As shown in the figure, the output signal of the mixer is an envelope signal, including the signal and the switch signal. The outside of the envelope is a sinusoidal signal with the same frequency, and the signal inside the envelope is at the same frequency as the square wave signal. The amplitude of the square wave signal in the figure is double V_{ref}.

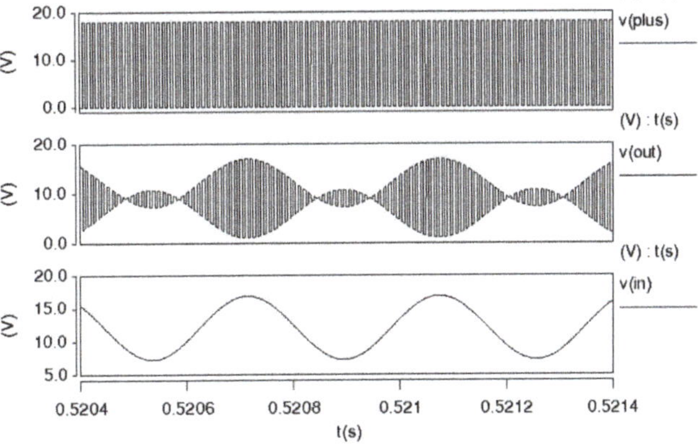

Figure 14. Simulated result of mixer.

The overall transient simulation results of the closed-loop drive system are shown in Figure 15. V(drive_sense_out) is the detection signal of the micromachined gyroscope in the driving direction, and its vibration speed is represented by V(speed). The signal of the driving loop based on frequency modulation reaches a stable state soon after the oscillation. The stabilization time is approximately 200 ms, the detection signal amplitude is about 21.32 mV and the amplitude does not change over time after stabilization. Therefore, the designed frequency-based demodulation circuit successfully stabilizes the vibration.

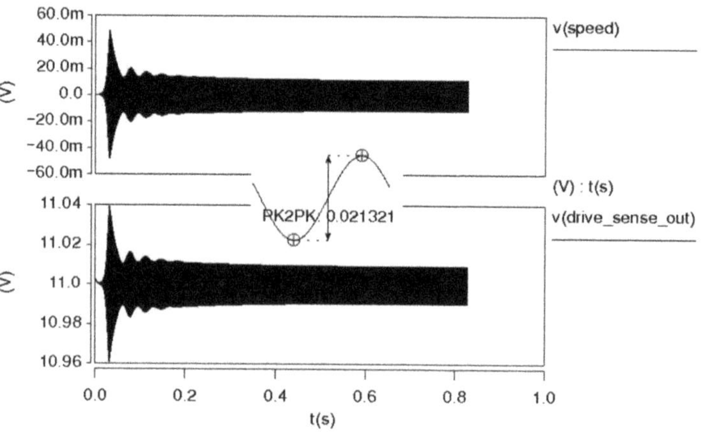

Figure 15. Transient response of closed-loop driving system.

The waveform diagram of the driving signal of the micromachined gyroscope in the steady state is shown in Figure 16. V(drive_ac+) and V(drive_ac-) with opposite phases are the frequency-modulated driving voltage signals loaded on the driving electrodes of the micromechanical gyroscope, respectively. V(drive_sense_out) is a drive detection signal. The detection signal frequency in the driving direction of the micromachined gyroscope is equal to its natural frequency, which is determined via its frequency selection characteristics. The stable average frequency is 4504 Hz and the frequency deviation is 1 Hz. Based on the mean square deviation of the simulation data after the system is stable, the frequency jitter is 0.0221 Hz.

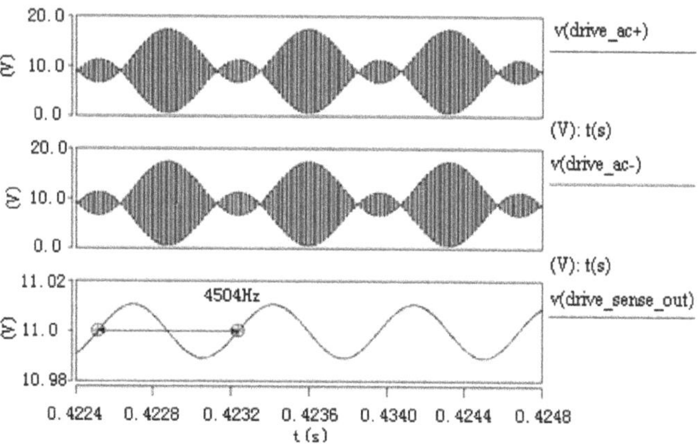

Figure 16. Transient response of micromachined gyroscope driving circuit.

5.3. Experimental Test

In order to verify the feasibility of the drive circuit design scheme of the MEMS gyroscope, the closed-loop self-excited oscillation of the gyroscope interface ASIC chip was tested experimentally. Figure 17a shows the test result of the electrostatic force drive modulation signal of the MEMS gyroscope. The drive force signal was frequency-modulated, which is consistent with the designed driving force signal. Figure 17b presents the time domain test chart of the MEMS gyroscope drive signal, and the test results show that the designed driving circuit can achieve the closed-loop driving function.

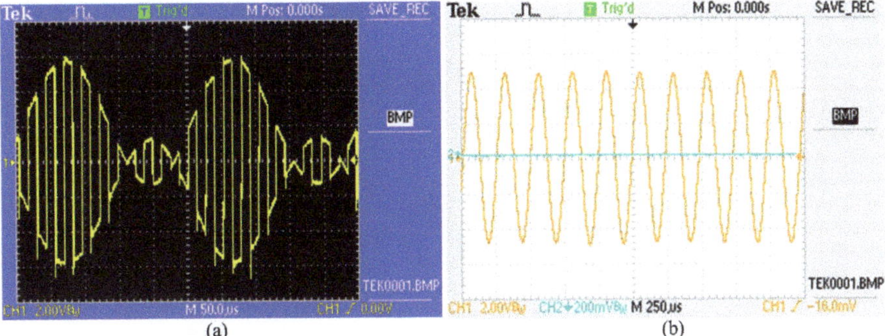

(a)

(b)

Figure 17. Experimental results of driving circuit of the MEMS gyroscope: (**a**) test results of electrostatic driving modulation signal; (**b**) test results of driving signal time domain.

Figure 18a shows the test results of the driving vibration signals of ASIC, the driving circuit interface of the MEMS gyroscope. One point was picked every 1 s, and one hundred points were picked, respectively, for each test experiment. The experimental results demonstrate that the interface ASIC can realize closed-loop self-excited driving with stable amplitude and frequency under normal pressure. Frequency stability was measured using the frequency channel of a multimeter (Agilent34410A). The resonant frequency was 4.504 kHz, the frequency fluctuation was less than 0.1 Hz and the relative stability was about 0.0025%. The stability of the driving displacement amplitude was tested using the voltage channel of a multimeter. The relative stability of the driving displacement amplitude was about 0.0050%.

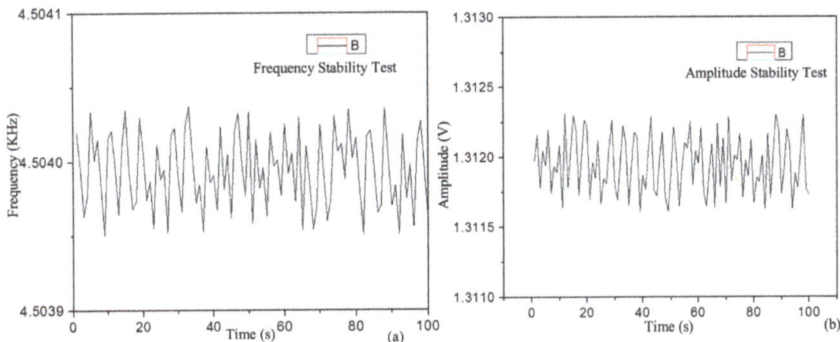

(a)

(b)

Figure 18. Test results of driving circuit stability of the MEMS gyroscope: (**a**) driving displacement amplitude stability; (**b**) drive signal frequency stability.

Table 2 compares the performance of various driving circuits for the MEMS gyroscope. The self-excited oscillation driving circuit of this design, which was based on the secondary demodulation principle, was monolithically integrated using 0.5 μm CMOS process technology, which has a high degree of integration. By comparing the measured results of the method of optimizing phase noise with the experimental results of the scheme without considering phase noise optimization, Table 2 shows that the self-excited oscillation driving circuit designed using optimized phase noise improved the driving stability. The MEMS gyroscope interface circuit designed in this work achieved monolithic integration and good overall performance.

Table 2. Comparison of the MEMS gyroscope's driving circuit stability.

Ref	Frequency (kHz)	Frequency Stability	Amplitude Stability
[26]	-	0.0290%	0.0470%
[27]	54.76	0.003%	0.0025%
This work	4.504	0.0025%	0.005%

6. Conclusions

In this paper, a self-excited oscillation driving interface circuit based on the principle of quadratic demodulation was proposed. This circuit is suitable for low-Q gyroscopes under normal temperature and pressure. A self-excited closed-loop driving circuit consisting of a phase shifter, PI controller and mixer was designed, and it can precisely and stably control the mechanical sensitive element. This interface circuit can eliminate the influence of the drive electrode on the displacement electrode and reduce the frequency jitter, improving the overall performance of the gyroscope. Thus, the gyro's driving interface ASIC is monolithically integrated using standard 0.5 μm CMOS process technology. Then, the system-level modeling and simulation of the gyroscope driving loop were carried out using SIMULINK, and the frequency and amplitude stability of the gyroscope driving circuit were tested and verified. The results show that the relative stability of the drive circuit frequency stability was about 0.0025%, and the frequency fluctuation was less than 0.1 Hz. The relative stability of the driving displacement amplitude was about 0.0050%. The MEMS gyroscope driving circuit test experiment verifies the correctness of the interface ASIC chip design scheme, and the gyroscope driving circuit can achieve a good, stable amplitude of self-excited oscillation.

Author Contributions: Conceptualization, T.H.; methodology, G.W. and C.D.; data curation, M.R.; writing—original draft preparation, T.H.; writing—review and editing, Z.Z.; supervision, X.J.; project administration, Z.Z. All authors have read and agreed to the published version of the manuscript.

Funding: This research was funded by Jinhua Science and Technology Bureau, grant numbers 2022-4-060, 2022-4-062 and 2022-4-237, and Jinhua Advanced Research Institute, grant numbers G202205, G202209 and G202207.

Data Availability Statement: The data that support the findings of this study are available from the corresponding author upon reasonable request.

Conflicts of Interest: The authors declare no conflict of interest.

References

1. Li, X.; Wang, W.; Wang, S.; Peng, Y.; Jin, X. Status and development trend of MEMS inertial sensors. *Meas. Technol.* **2019**, *39*, 7. [CrossRef]
2. Yi, J.; Jiang, N.; Zhuang, S.; Guo, S.; Zhang, J. Status and development of MEMS solid-state fluctuating gyro resonators. *Micro Nanoelectron.* **2018**, *55*, 738–745.
3. Yang, B.; Wang, S.R.; Li, K.Y.; Zhu, X.; Cao, H. Silicon tuning gyroscope tuned by negative stiffness effect. *Opt. Precis. Eng.* **2010**, *18*, 2398–2406.
4. Hou, Z.; Kuang, Y.; Ou, F.; Xu, Q.; Miao, T.; Xiao, D.; Wu, X. A quadrature compensation method to improve the performance of the butterfly vibratory gyroscope. *Sens. Actuators A Phys.* **2021**, *319*, 112527. [CrossRef]
5. Jeong, C.; Seok, S.; Lee, B.; Kim, H.; Chun, K. A study on resonant frequency and Q factor tunings for MEMS vibratory gyroscopes. *J. Micromech. Microeng.* **2004**, *14*, 1530–1536. [CrossRef]

6. Miao, T.; Hu, X.; Zhou, X.; Wu, X.; Hou, Z.; Xiao, D. A Million-order Effective Quality Factor MEMS Resonator by Mechanical Pumping. In Proceedings of the 2020 IEEE International Symposium on Inertial Sensors and Systems (INERTIAL), Hiroshima, Japan, 23–26 March 2020; pp. 1–4.

7. Xu, P.; Si, C.; He, Y.; Wei, Z.; Jia, L.; Han, G.; Ning, J.; Yang, F. A Novel High-Q Dual-Mass MEMS Tuning Fork Gyroscope Based on 3D Wafer-Level Packaging. *Sensors* **2021**, *21*, 6428. [CrossRef]

8. Xu, P.; Wei, Z.; Guo, Z.; Jia, L.; Han, G.; Si, C.; Ning, J.; Yang, F. A Real-Time Circuit Phase Delay Correction System for MEMS Vibratory Gyroscopes. *Micromachines* **2021**, *12*, 506. [CrossRef]

9. Keymeulen, D.; Peay, C.; Foor, D.; Trung, T.; Bakhshi, A.; Withington, P.; Yee, K.; Terrile, R. Control of MEMS Disc Resonance Gyroscope (DRG) using an FPGA Platform. In Proceedings of the Aerospace Conference, Big Sky, MT, USA, 1 March 2008.

10. Goto, K.; Harada, S.; Hata, Y.; Ito, K.; Wado, H.; Cho, J.Y.; Najafi, K. High Q-Factor Mode-Matched Silicon Gyroscope with a Ladder Structure. In Proceedings of the 2020 IEEE International Symposium on Inertial Sensors and Systems (INERTIAL), Hiroshima, Japan, 23–26 March 2020.

11. Elwell, J.M. Micromechanical Inertial Swors for Commercial and Military Applications. In Proceedings of the 50th Annual Meeting of The Institute of Navigation, Washington, DC, USA, 11–13 May 1994.

12. Joachim, D.; Lin, L. Selective polysilicon deposition for frequency tuning of MEMS resonators. In Proceedings of the Micro Electro Mechanical Systems, 2002, the Fifteenth IEEE International Conference on IEEE, Las Vegas, NV, USA, 24 January 2002.

13. Wang, Z.; Fei, J. Fractional-Order Terminal Sliding-Mode Control Using Self-Evolving Recurrent Chebyshev Fuzzy Neural Network for MEMS Gyroscope. *IEEE Trans. Fuzzy Syst.* **2022**, *30*, 2747–2758. [CrossRef]

14. Lv, R.S.; Fu, Q.; Yin, L.; Gao, Y.; Bai, W.; Zhang, W.B.; Zhang, Y.F.; Chen, W.P.; Liu, X.W. An Interface ASIC for MEMS Vibratory Gyroscopes with Nonlinear Driving Control. *Micromachines* **2019**, *10*, 270. [CrossRef]

15. Li, X.Y.; Hu, J.P.; Liu, X.W. A high-performance digital interface circuit for a high-Q micro-electromechanical system accelerometer. *Micromachines* **2018**, *9*, 675. [CrossRef]

16. Huang, F.; Liang, Y. Analysis and design of the system of a total digital Si-gyroscope. *Int. J. Mod. Phys. B* **2017**, *31*, 1741008. [CrossRef]

17. Ahn, C.H.; Ng, E.J.; Hong, V.A.; Yang, Y.S.; Lee, B.J.; Flader, I.; Kenny, T.W. Mode-matching of wineglass mode disk resonator gyroscope in (100) single crystal silicon. *J. Microelectromech. Syst.* **2015**, *24*, 343–350. [CrossRef]

18. Sonmezoglu, S.; Alper, S.E.; Akin, T. An automatically mode-matched MEMS gyroscope with wide and tunable bandwidth. *J. Microelectromech. Syst.* **2014**, *23*, 284–297. [CrossRef]

19. Balachandran, G.K.; Petkov, V.P.; Mayer, T.; Balslink, T. A 3-axis gyroscope for electronic stability control with continuous self-test. *IEEE J. Solid-State Circuits* **2016**, *51*, 177–186.

20. Lv, R.S.; Chen, W.P.; Yin, L.; Fu, Q.; Liu, X.W.; Yan, J.M. A closed-loop SD modulator for micromechanical capacitive sensors. *IEICE Electron. Express* **2018**, *15*, 20171112. [CrossRef]

21. Lv, R.S.; Chen, W.P.; Liu, X.W. A high-dynamic-range switched-capacitor sigma-delta ADC for digital micromechanical vibration gyroscopes. *Micromachines* **2018**, *9*, 372. [CrossRef]

22. Zhao, Y.; Zhao, J.; Wang, X.; Xia, G.M.; Shi, Q.; Qiu, A.P.; Xu, Y.P. A sub-0.1 degrees/h bias-instability split-mode MEMS gyroscope with CMOS readout circuit. *IEEE J. Solid-State Circuits* **2018**, *53*, 2636–2650. [CrossRef]

23. Feng, C.; Liu, D.; Ma, H.; Qing, C.; Li, H.; Feng, L. Design, fabrication and test of transmissive Si_3N_4 waveguide ring resonator. *IEEE Sens. J.* **2021**, *21*, 22918–22926. [CrossRef]

24. Yang, C.; Li, H.S.; Xu, L.; Zhu, K. Low-frequency modulated excitation-based automatic mode-matching technique for silicon micro gyroscope. *Chin. J. Inert. Technol.* **2016**, *1*, 542–547.

25. Bu, F.; Peng, Y.; Xu, D.; Zhao, H. A MEMS gyro mode-matching method combining quadrature control and phase detection. *Chin. J. Inert. Technol.* **2018**, *2*, 470–477.

26. Zhixiong, Z.; Lihui, F.; Yu-nan, S. Temperature Modeling and Compensation of Double h Quartz Tuning Fork Gyroscope. *Procedia Eng.* **2011**, *15*, 752–756. [CrossRef]

27. Zhang, H.; Chen, W.P.; Liang, Y. An Interface ASIC Design of MEMS Gyroscope with Analog Closed Loop Driving. *Sensors* **2023**, *23*, 2615. [CrossRef] [PubMed]

 *micromachines*

Article

High-G MEMS Accelerometer Calibration Denoising Method Based on EMD and Time-Frequency Peak Filtering

Chenguang Wang [1,2], Yuchen Cui [2,3], Yang Liu [4], Ke Li [2,3,*] and Chong Shen [2,3,*]

[1] School of Information and Communication Engineering, North University of China, Taiyuan 030051, China; wangchenguang@nuc.edu.cn
[2] Science and Technology on Electronic Test & Measurement Laboratory, North University of China, Taiyuan 030051, China; 2006040621@st.nuc.edu.cn
[3] School of Instrument and Electronics, North University of China, Taiyuan 030051, China
[4] Shanxi North Machine-Building Co., Ltd., Taiyuan 030051, China; m18734913028@163.com
[*] Correspondence: nuclike@outlook.com (K.L.); shenchong@nuc.edu.cn (C.S.); Tel.: +86-195-1037-7595 (K.L.)

Abstract: In order to remove noise generated during the accelerometer calibration process, an accelerometer denoising method based on empirical mode decomposition (EMD) and time-frequency peak filtering (TFPF) is proposed in this paper. Firstly, a new design of the accelerometer structure is introduced and analyzed by finite element analysis software. Then, an algorithm combining EMD and TFPF is proposed for the first time to deal with the noise of the accelerometer calibration process. Specific steps taken are to remove the intrinsic mode function (IMF) component of the high frequency band after the EMD decomposition, and then to use the TFPF algorithm to process the IMF component of the medium frequency band; meanwhile, the IMF component of the low frequency band is reserved, and finally the signal is reconstructed. The reconstruction results show that the algorithm can effectively suppress the random noise generated during the calibration process. The results of spectrum analysis show that EMD + TFPF can effectively protect the characteristics of the original signal and that the error can be controlled within 0.5%. Finally, Allan variance is used to analyze the results of the three methods to verify the filtering effect. The results show that the filtering effect of EMD + TFPF is the most obvious, being 97.4% higher than the original data.

Keywords: MEMS accelerometer; empirical mode decomposition; time-frequency peak filtering; high-g calibration

Citation: Wang, C.; Cui, Y.; Liu, Y.; Li, K.; Shen, C. High-G MEMS Accelerometer Calibration Denoising Method Based on EMD and Time-Frequency Peak Filtering. *Micromachines* **2023**, *14*, 970. https://doi.org/10.3390/mi14050970

Academic Editor: Ha Duong Ngo

Received: 13 April 2023
Revised: 27 April 2023
Accepted: 27 April 2023
Published: 28 April 2023

1. Introduction

Inertial technology is a cross-integrated technology involving inertial navigation [1] and guidance [2], inertial systems [3] and related disciplines [4–6]. Accelerometers, as the most basic inertial devices, are widely used in automobile, defense and impact measurement [7,8]. However, the requirements for accuracy are becoming higher and higher. Therefore, how to improve the accuracy of the accelerometer calibration process is very important. Many researchers have undertaken a lot of work to improve the accuracy of calibration. The approaches pursued include those described below.

Accelerometer denoising methods can be divided into the following two types: hardware-based methods and software methods. Zhang proposed a capacitive micro-electro-mechanical system (MEMS) accelerometer based on an asymmetrical anti-spring structure, which has high sensitivity and low noise performance. The accelerometer has been successfully applied in an earthquake [9]. Rao developed a micromachined micro-g capacitive accelerometer with a silicon-based spring-mass sensing element. By increasing the weight of the proof mass, the thermal noise was reduced [10]. Kamada devised a sensor architecture with a unique perforated and electrode-separated mass structure; the accelerometer achieved both low noise and low power consumption [11]. Yeh presented a monolithically integrated CMOS-MEMS three-axis capacitive accelerometer, which achieved the goal of low noise and low zero gravity [12]. Utz designed an accelerometer consisting of an ultra-low noise CMOS integrated readout-IC and a high-precision bulk micro-machined sensing element; the acceleration equivalent noise was significantly reduced [13]. Liu presented a low-power micromechanical capacitive accelerometer system with hybrid signal output to ensure a low noise output signal [14]. Najafi introduced classic and knowledge-based intelligent controllers for regulation of a vibratory MEMS accelerometer to filter stochastic noises [15]. Edalafar developed a high-performance micromachined capacitive accelerometer with the feature of low noise [16].

Another denoising method is that of software compensation. Algorithm compensation is more convenient and flexible than hardware-based compensation. Yan proposed a hybrid denoising algorithm based on time-frequency peak filtering (TFPF), local mean decomposition (LMD) and sample entropy (SE) to decrease the influence of noise on the high-g MEMS accelerometer (HGMA) output signal [17]. Guo presented a Kalman filtering method based on information fusion. By using the MEMS gyroscope and line accelerometer signals to implement the filtering function under the Kalman algorithm, the noise of the gyroscope signal was significantly reduced [18]. Zou combined ensemble empirical mode decomposition (EEMD) with time-domain integration. The fusion algorithm was based on a bridge dynamic displacement reconstruction method, then the high-frequency ambient noise was effectively eliminated [19]. Ding combined a wavelet de-noising method with a nonlinear independent component analysis (ICA) method to tackle the nonlinear BSS problem with additive noise [20]. Shen studied a novel multiple inputs/single output model based on a genetic algorithm (GA) and the Elman neural network (Elman NN) to improve the temperature drift modeling precision of gyroscopes [21]. Jiang adopted an artificial intelligence (AI) method to de-noise the MEMS inertial measurement unit (IMU) output signals [22]. Lu proposed a denoising method based on the combination of empirical mode decomposition (EMD) and wavelet threshold [23]. Zhu proposed a radial basis function (RBF) neural network (NN) + genetic algorithm (GA) + Kalman filter (KF) method, combined with a temperature drift model, so that noise characteristics were well optimized [24]. Zhang proposed a dual modulation method. The intensity modulation enabled movement of the signal to a high frequency, and the light source noise was suppressed perfectly by combining phase modulation [25]. Mokhtari implemented wavelet de-noising and de-trending techniques in order to filter angular accelerations [26]. He proposed the Kalman filter method to preprocess the raw data to reduce noise [27]. Abbasi-Kesbi presented a method based on the L2-norm total variation (LTV) algorithm. The obtained signals from the accelerometer were denoised [28]. Kou designed a hybrid algorithm based on forward linear prediction (FLP), a grey accumulated generating operation (AGO) and lifting wavelet transform (LWT) to achieve a better denoising effect [29]. Shen proposed a parallel processing algorithm, which was based on variational mode decomposition (VMD) and an augmented nonlinear differentiator (AND), to improve the effectiveness of the denoising process [30]. Wang studied a method based on an improved ensemble local mean decomposition to reduce noise [31]. Cao proposed a structural equivalent electronic model and improved differential interface based on weak signal detection technology to improve the accuracy of inertial devices [32]. Wang used a method based on ensemble empirical mode decomposition (EEMD) and multipoint optimal minimum entropy deconvolution adjusted (MOMEDA) to reduce noise [33]. Cao introduced a sense mode closed-loop method for a MEMS gyroscope, which was based on the dipole temperature compensation method, and improved accuracy effectively [34]. Cai proposed a parallel processing model for eliminating gyroscope noise and temperature drift based on multi-objective particle swarm optimization based on a variational modal decomposition-time-frequency peak filter (MOVMD–TFPF) and the beetle antennae search algorithm– Elman neural network (BAS–Elman NN) [35]. Shen introduced a noise reduction algorithm based on an improved empirical mode decomposition (EMD) and forward linear prediction (FLP) to reduce the noise of a fiber optic gyroscope; the standard deviation of the gyroscope output signal was effectively reduced [36]. Cao proposed an IAWTD (improved adaptive wavelet threshold de-noising)-CSVM (C-means support vector machine)-EEMD (ensemble empirical mode decomposition) algorithm to compensate for the humidity drift of a gyroscope; this algorithm effectively reduced the quantization noise, bias stability and angle random walk of a MEMS gyroscope [37]. Ma introduced a fusion algorithm based on an immune-based particle swarm optimization (IPSO) improved VMD and BP neural network to reduce the temperature drift and output signal noise of the gyroscope [38]. Cao proposed a novel compensation method based on a permutation entropy local characteristic-scale decomposition (PE-LCD) and adaptive network-based fuzzy inference system (ANFIS) for a dual-mass MEMS gyroscope [39]. Ma introduced a parallel denoising model for a dual-mass MEMS gyroscope based on PE (Permutation entropy)-ITD (Intrinsic time scale decomposition) and SA (Simulated annealing)-ELM (Extreme learning machine) [40].

Although the above research institutions have been able to approximate the nonlinear relationship of the error using a variety of algorithm combinations, no significant suppression effect on the noise generated during the accelerometer calibration process has been demonstrated, and the characteristics of the original signal cannot be effectively preserved. To address the above problems, this paper proposes an accelerometer calibration process denoising algorithm based on empirical mode decomposition (EMD) and time-frequency peak filtering (TFPF). First, the intrinsic mode function (IMF) components of the high-frequency band after EMD decomposition are removed. Then, the TFPF algorithm is used to process the IMF components of the intermediate frequency, while the low-frequency IMF components are retained. Finally, the signal is reconstructed. The spectrum analysis results show that EMD + TFPF can effectively protect the characteristics of the original signal, and the error can be controlled within 0.5%. The results of Allan variance analysis show that the filtering effect is improved by 97.4% compared with the original data.

This paper introduces the proposed method as follows: Section 2 introduces the structure of the high g-value accelerometer, while Section 3 describes the construction of the EMD and TFPF fusion algorithm. The impact experiments performed on the accelerometer and signal processing are described in Section 4. Finally, Section 5 presents the conclusions of the paper.

2. Structure of MEMS Accelerometer and Working Mode Analysis

The original signal collected in this paper is from a newly designed and manufactured high-g MEMS accelerometer (HGMA) [41,42]. It is a kind of accelerometer with a high impact survival rate and high measuring range. The detection method for the HGMA used is the piezo resistance and the output signal is the voltage. The HGMA adopts a four beam and island structure. The frame, four beams and the center mass are all rectangular, which is conducive to processing. The structure diagram and parameters are shown in Figure 1.

The coordinate system is constructed with the cross-section of the accelerometer. The central dividing line of the cross-section is the z-axis (specifying that the direction is positive to the downward direction). The other middle line is the x-axis, and the right direction is positive. The frame constructed is shown in Figure 1. The beam length, width and thickness are a_1, b_1 and c_1, respectively; the mass length, width and thickness are $2a_2$, b_2 and c_2, respectively. The size values are shown in Table 1.

The accelerometer is simulated and analyzed through ANSYS soft at four primary resonance modes; Figure 2a–d show the first, second, third and fourth modes, respectively. The first mode mass moves along the z-axis and is the working mode; the second mode mass rotates around the x-axis; the third mode mass rotates around the y-axis; the fourth mode mass and frame move along the z-axis. The resonant frequencies of the four modes are shown in Table 2, which indicates that the 1st order is the working mode of HGMA and its resonant frequency is 408 kHz. The 2nd-order mode resonant frequency is 667 kHz and has a 260 kHz gap with the 1st-order mode, which means that the coupling movement between these two modes is tiny and is good for HGMA linearity.

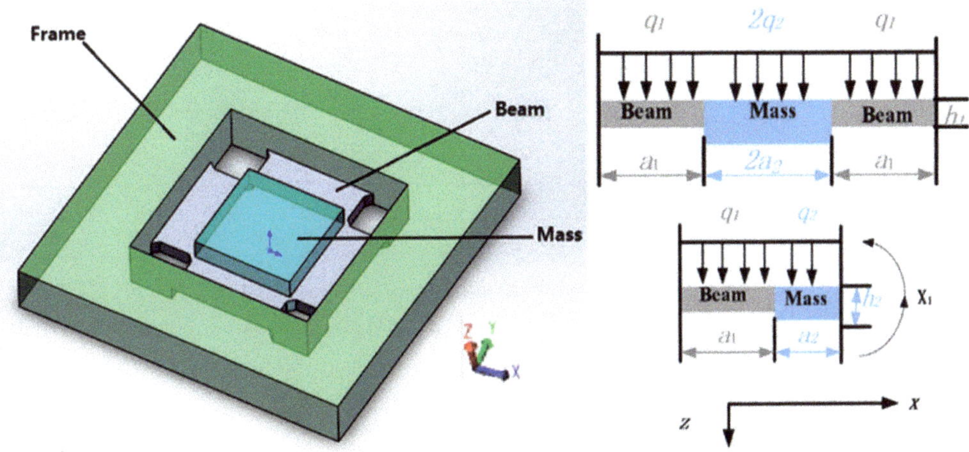

Figure 1. HGMA structure schematic and size.

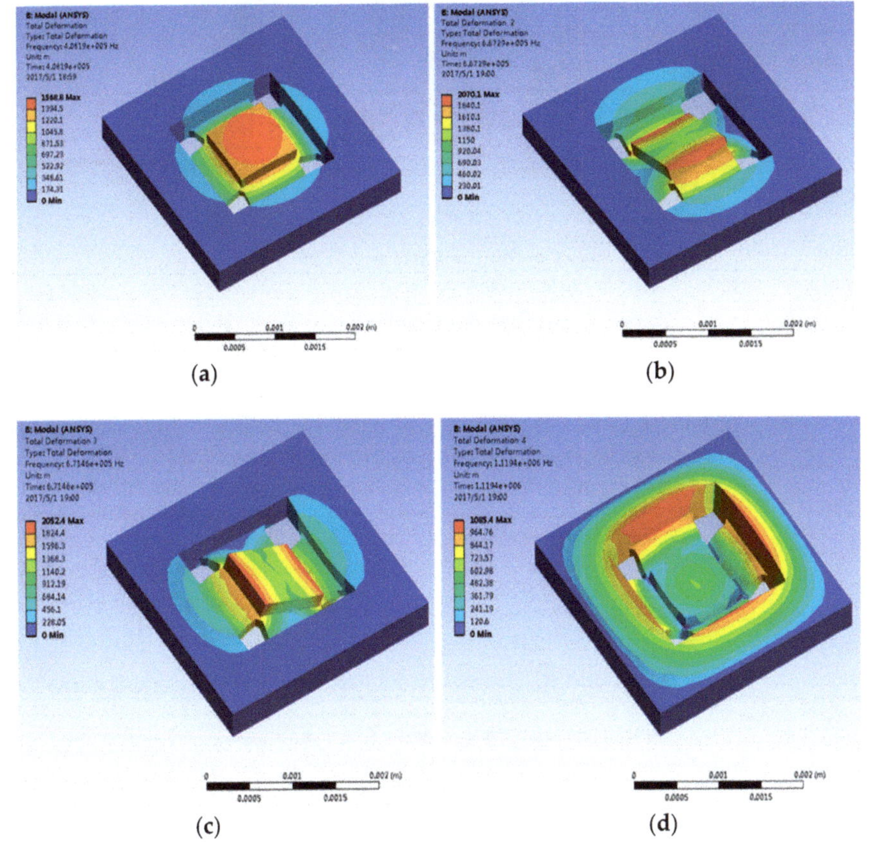

Figure 2. Mode simulation of HGMA structure. (**a**–**d**) are the 1st, 2nd, 3rd and 4th order modes.

Table 1. Structural parameters of the HGMA.

Parameters	Beam			Mass		
	length (a_1)	width (b_1)	height (c_1)	length (a_2)	width (b_2)	height (c_1)
size/μm	350	800	80	800	800	200

Table 2. Resonant frequencies of the four modes.

Mode Shapes	1	2	3	4
Resonant Frequency /kHz	408	667	671	1119

The structure of the HGMA is made of silicon bonding on glass. The main technological process is divided into 12 steps. SEM photos and CCD photos of the accelerometer structure are shown in Figure 3.

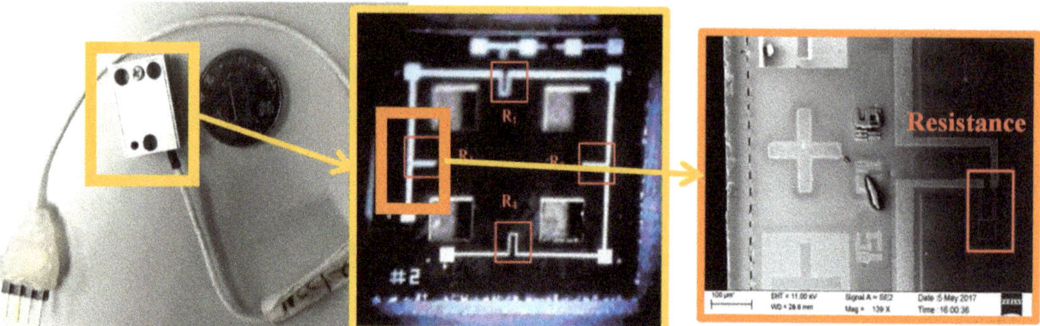

Figure 3. Overall photo, CCD photo and SEM photo of HGMA.

3. Methodology Combining EMD and TFPF

3.1. Empirical Mode Decomposition

Empirical mode decomposition (EMD) is an effective method for processing non-linear and non-stationary time-varying sequences. The method can adaptively decompose signals based on the time-scale characteristics of the data itself. The algorithm can decompose complicated time-series data into a finite number of intrinsic mode functions (IMF). The IMF component has all the fluctuation information of the original data at the corresponding time-scale. The specific steps of the EMD are as follows [43]:

Step 1: Find all the maximum points of the original signal $x(t)$ and use the cubic spline interpolation function to form the upper envelope of the original data. Similarly, find all the minimum points and use the same interpolation function to form the lower envelope. The average of the upper and lower envelopes is set as m_1; its value is shown in Equation (1).

$$m_1 = \frac{1}{2}(x_{\min}(t) + x_{\max}(t)) \tag{1}$$

where $x_{\min}$ and $x_{\max}$ represent the maximum envelope and the minimum envelope, respectively.

Step 2: Subtract the average value from the original signal to obtain a new high-frequency sequence with the low frequencies removed, denoted as $h_1(t)$.

$$h_1(t) = x(t) - m_1 \tag{2}$$

where $x(t)$ is the original signal.

Step 3: If $h_1(t)$ is not an IMF component, set $h_1(t)$ as the original data, repeat the above process.

Continue with Step 1, obtain the average value m_{11}.

$$m_{11} = \frac{1}{2}(h_{1\min}(t) + h_{1\max}(t)) \tag{3}$$

Continue with Step 2, obtain $h_{11}(t)$.

$$h_{11}(t) = h_1(t) - m_{11} \tag{4}$$

If $h_{11}(t)$ is still not an IMF component, repeat the loop k times and obtain m_{1k} and $h_{1k}(t)$ until $h_{1k}(t)$ meets the conditions of the IMF. The results are as follows:

$$m_{1k} = \frac{1}{2}(h_{1k\min}(t) + h_{1k\max}(t)) \tag{5}$$

$$h_{1k}(t) = h_{1k-1}(t) - m_{1k} \tag{6}$$

Step 4: Assume $c_1(t)$ as the highest frequency component of the original signal, set the value as follows:

$$c_1(t) = h_{1k}(t) \tag{7}$$

Step 5: Separate $c_1(t)$ from $x(t)$ and obtain Equation (8).

$$x_1(t) = x(t) - c_1(t) \tag{8}$$

where $x_1(t)$ is the new high-frequency sequence with low frequency removed.

Step 6: The value obtained by Equation (8) is subjected to the above sieving process, so the second IMF component can be obtained. After repeating n times, obtaining the nth IMF-compliant component, namely:

$$\begin{cases} x_1(t) - c_2(t) = x_2(t) \\ \quad \cdots \\ x_{n-1}(t) - c_n(t) = x_n(t) \end{cases} \tag{9}$$

Step 7: If the last component $x_n(t)$ is a monotonic function and its value is small or tends to zero, the decomposition is terminated and the steps are not repeated. Finally, the relationship between the original signal and each component is shown in Equation (10).

$$x(t) = \sum_{i=1}^{n} c_i(t) + x_n(t) \tag{10}$$

where $c_1(t)$, $c_2(t)$, $c_3(t)$, ..., $c_n(t)$ are the individual IMF components; $x_n(t)$ represents the average trend of the signal.

3.2. Time-Frequency Peak Filtering

For the noise filtering problem of low SNR (signal-to-noise ratio) signals, the time-frequency peak filtering (TFPF) filter works well. In this method, the noisy signal is coded and modulated into a certain analytical signal frequency, and the Wigner–Ville time-frequency distribution is used to obtain an estimate of the peak frequency of the analyzed signal. According to the characteristics of the modulation signal of the noise in the Wigner–Ville distribution, when the time-frequency peak is extracted, the influence can be filtered out; finally, the analytical signal is restored, and the signal noise reduction can be realized.

The time-frequency peak filtering method is very effective for the extraction of weak signals. It can be mainly divided into the following steps [44–46]:

Step 1: Assume that the noisy signal is as follows:

$$s_c(t) = s(t) + c(t) \tag{11}$$

where $s_c(t)$ is the signal including noise, $s(t)$ is the signal and $c(t)$ is the Gaussian white noise.

Step 2: Encode the signal and obtain Formula (12).

$$z(t) = e^{j2\pi\mu \int_{-\infty}^{t} s_c(\lambda)d\lambda} \tag{12}$$

where $z(t)$ is the analytical signal and μ is the frequency modulation factor with a value range from 0–1.

The discrete expression of Formula (12) is as follows:

$$z(n) = e^{j2\pi\mu \sum_{i=0}^{n} s_c(iT_s)T_s} \tag{13}$$

where T_s is the sampling time.

According to the sampling theorem, the highest frequency of the sampling signal is $f_s/2$ and the time domain discrete signal has no negative frequency. Then, set the instantaneous frequency of $z(n)$ as follows:

$$f_i(n) = \frac{\left[\sum_{i=0}^{n} s_c(iT_s) - \sum_{i=0}^{n-1} s_c(iT_s)T_s\right]}{T_s} = s_c(iT_s) \tag{14}$$

where $f_i(n)$ is the instantaneous frequency of $z(n)$ with a frequency range between 0 and $f_s/2$.

From the above analysis, Formula (15) can be obtained.

$$0 \leq s_c(iT_s) \leq f_s/2 \tag{15}$$

In general, for convenience of calculation, Formula (13) can be changed to the following:

$$z(n) = e^{j2\pi\mu \sum_{i=0}^{n} s_c(i)} \tag{16}$$

From Equation (16), Equation (15) can be changed to the following:

$$0 \leq s_c(i) \leq 1/2 \tag{17}$$

It can be derived from Formula (17) that the signal amplitude should be limited within the range and that there will be no overlap when it is modulated into a frequency. In order to avoid signal distortion during frequency modulation, scale the acquired signal [47]:

$$s_c(i) = (a - b)\frac{s_c(i) - \min[s_c(i)]}{\max[s_c(i)] - \min[s_c(i)]} + b \tag{18}$$

where $b \leq a$, and their values range from 0 to 0.5.

Step 3: Find the Wigner–Ville distribution (WVD) of the resolved signal [48].

$$W_{z_x}(t, f) = \int_{-\infty}^{\infty} z_x(t + \tfrac{\tau}{2})z_x^*(t - \tfrac{\tau}{2})e^{j2\pi f\tau}d\tau \tag{19}$$

where t represents time, τ represents the integral variable, and f represents the frequency.

Since the actual signal has nonlinear characteristics, the pseudo-Wigner–Ville distribution (PWVD) is used to find the time-frequency distribution of the signal to ensure that the estimation of the signal in the window is unbiased [46].

The PWVD distribution of the noisy signals is as follows:

$$W_{pz_x}(t, f) = \int_{-\infty}^{\infty} h(\tau)z_x(t + \frac{\tau}{2})z_x^*(t - \frac{\tau}{2})e^{j2\pi f\tau}d\tau \tag{20}$$

where $h(\tau)$ is a real-valued window function.

The discrete expression of Formula (20) is as follows:

$$W_{pz_x}(n, m) = \sum_{l=-L}^{L} h(l) z_x(n + l) z_x{}^*(n - l) e^{j2\pi mn} \tag{21}$$

where $h(l)$ is a window function, and its width is 2L + 1.

The maximum value of the time-frequency distribution of the analytical signal is taken as the estimate of the instantaneous frequency by the frequency variable, which is shown in Formula (22).

$$\hat{f}_z = \frac{1}{\mu}\arg_f\max\left[W_{pz_x}(t, f)\right] \tag{22}$$

Step 4: Signal reduction

The instantaneous frequency is the estimate of the effective signal of the original signal. Set the estimate of the effective signal of the original signal as follows:

$$\hat{S}'(t) = \hat{f}_z \tag{23}$$

Perform an inverse scaling of the estimate of the effective signal and obtain the Formula (24) [47]:

$$\hat{S}(i) = \frac{(\hat{S}'(i) - b)(\max(s_c(i)) - \min s_c[(i)])}{(a - b)} + \min[s_c(i)] \tag{24}$$

If, when filtering, the effect (such as the signal-to-noise ratio) is not ideal, step 1 can be returned to for iterative filtering.

3.3. The EMD-TFPF Fusion Algorithm

The TFPF method can suppress part of the random noise. However, the TFPF method may not have a very good effect. However, the EMD algorithm can also filter but can lead to serious signal distortion. This section proposes a fusion algorithm based on empirical mode decomposition (EMD) and time-frequency peak filtering (TFPF).

Firstly, the EMD algorithm is used to decompose the acceleration calibration signal. The processed signal is divided into many IMF components, which are distributed from high frequency to low frequency. The decomposed IMF component is mainly divided into three parts: a noise part, a mixed signal and a trend term. There is almost no useful signal component in the high-frequency signal, so it can be ignored, which can effectively suppress part of the random noise. Then, we select a small amount of IMF for signal reconstruction. The TFPF algorithm is then implemented for each selected IMF to further filter the random signals generated in the calibration process; the specific approach uses a small window PWVD to suppress the residual weak random noise in the time-frequency analysis. Finally, the required signals are restored by adding the processed IMFs. The specific process is shown in Figure 4.

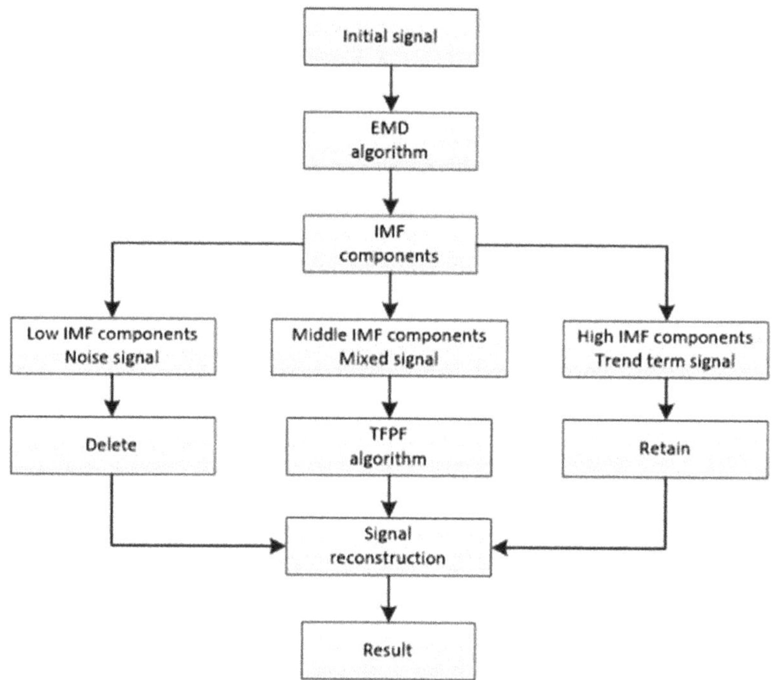

Figure 4. The process of the EMD-TFPF fusion algorithm.

The selection of IMF components is an important problem. In the calibration process of the accelerometer, there are two frequencies that need to be considered. The first is the shock frequency in the calibration process, and the second is the vibration frequency of the accelerometer. For the selected IMF components, spectrum analysis of each IMF component shows that there will be peaks around the shock frequency and vibration frequency. In order to ensure that the calibration process of the MEMS accelerometer does not produce spectrum distortion, the spectrum characteristics of the original signal should be retained as far as possible. Therefore, for the selection of these IMF components, the IMF component whose spectral energy ratio is greater than the average spectral energy of the IMF component is selected here.

EMD-TFPF is an improvement of the two filtering algorithms EMD and TFPF. The EMD-TFPF algorithm can effectively separate the signal noise and process each part of the IMF component accordingly. It avoids the problem that the real signal is obtained by EMD decomposition but the noise cannot be effectively filtered. It also avoids the problem that the window selection in TFPF is too large and affects the characteristics of the original signal.

4. Experiment and Signal Processing

As shown in Figure 5, the Hopkinson bar calibration system included a recycling box, deformeter, computer, Hopkinson bar, and compressed air. The power supply provided a +5 V voltage to the HGMA, the temperature was maintained at 25 °C (room temperature value), and the sampling rate was set to 20 MHz.

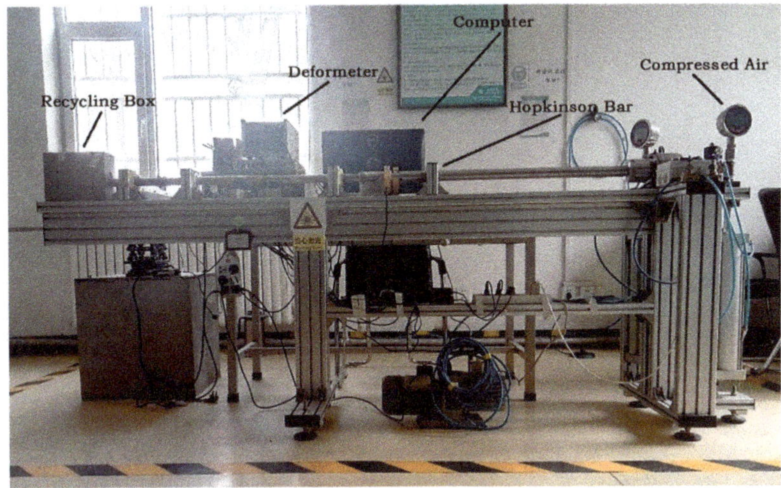

Figure 5. Hopkinson bar calibration system.

HGMA was calibrated in the Hopkinson bar calibration system. The HGMA output signal was acquired using a high-speed data acquisition system and a computer. The results are shown in Figure 6.

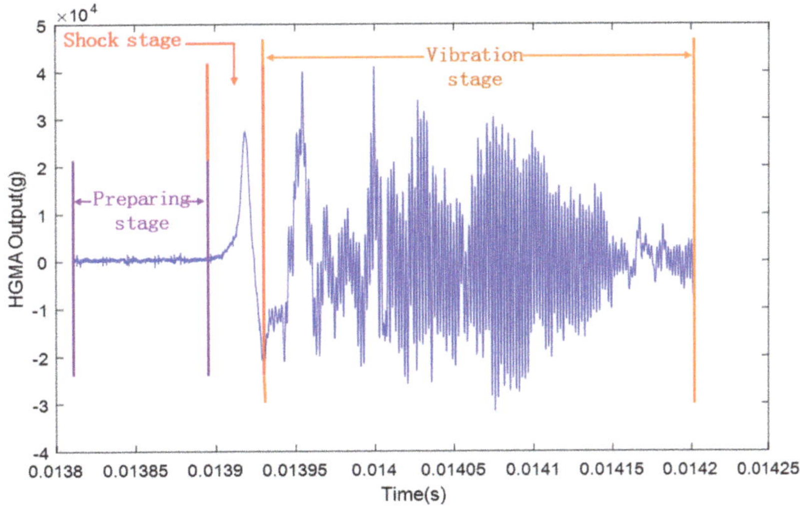

Figure 6. The output of the MEMS accelerometer in the calibration experiment.

It is worth noting that the HGMA calibration process is mainly divided into three phases, namely the preparation phase, the impact phase and the oscillation phase.

1. Preparation stage: this part mainly includes the bias characteristics of the HGMA. The noise signal is included. The maximum peak noise is approximately 1000 g.
2. Shock stage: this stage is the main part of the accelerometer calibration experiment; the output peak value is about 28,030 g and the pulse width is about 10 μs.
3. Vibration stage: this part mainly captures the vibration information for the HGMA, reflecting the dynamic characteristics of the HGMA.

The accelerometer calibration signal is decomposed by the EMD algorithm, as shown in Figure 7. It can be seen that the acceleration signal is finally decomposed into eight IMF components and a residual component. The IMF components are distributed from high frequency to low frequency.

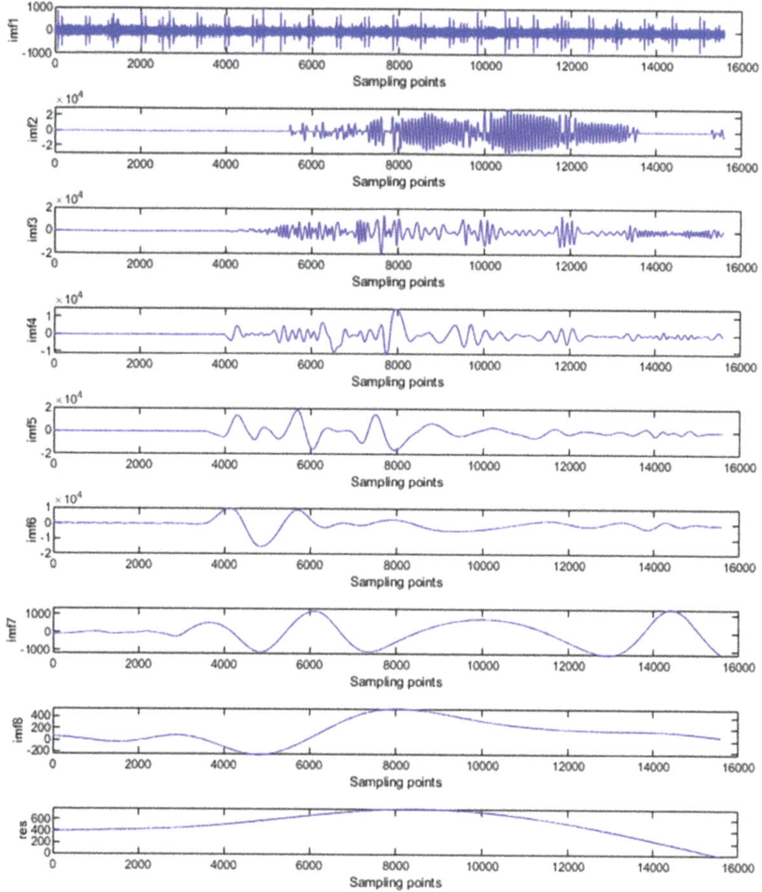

Figure 7. EMD decomposition of original signal.

According to the EMD low-pass filtering de-noising processing algorithm, the high-frequency part of the IMF component is removed directly, and the medium- and low-frequency IMF components are processed by Section 3.3. Therefore, each IMF component is finally layered. IMF1 is the noise term, while IMF2, IMF3, and IMF4 are mixed items, and IMF5–IMF8 are the trend items of the signal and should be reserved.

The mixed terms are processed by the TFPF algorithm, as shown in Figure 8. By observing the time series, it can be found that the TFPF algorithm can effectively suppress the random noise based on protection of the useful signal.

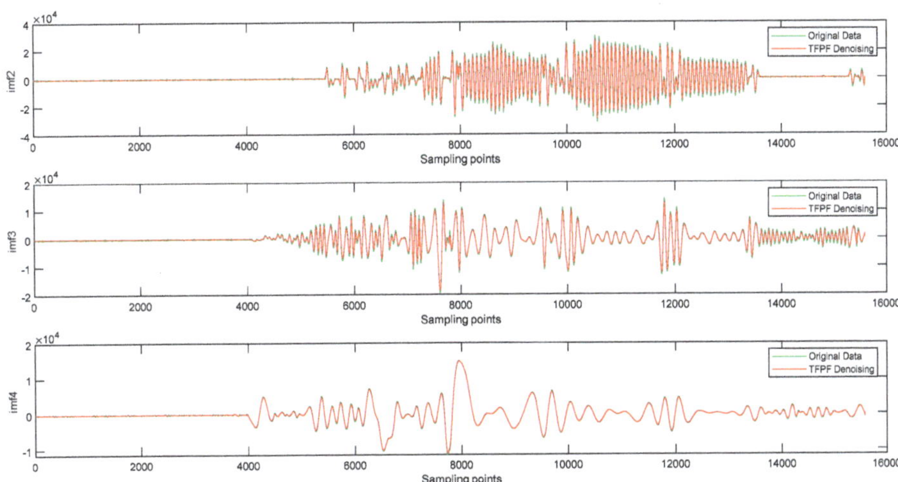

Figure 8. The results of mixed components after TFPF de-noising.

After the above signal processing, the accelerometer signal is reconstructed according to the steps in Figure 4. The results of reconstructing the accelerometer signal using the three methods are shown in Figure 8. It can be seen from Figure 8 that EMD has the best effect in the three stages.

In the preparation stage, the maximum noise of the original signal, the EMD denoising signal, the TFPF denoising signal, and the EMD + TFPF denoising signal were 1958 g, 878.7 g, 1483 g, 811.6 g, respectively. According to the results, among the three methods, the TFPF method had the worst denoising effect, and the EMD + TFPF method had the most obvious denoising effect.

In the shock stage, the curves of the original signal, the EMD denoising signal, the TFPF denoising signal, and the EMD + TFPF denoising signal almost coincided. The results show that the three algorithms contained the same information as the original signal.

In the vibration stage, the TFPF and EMD + TFPF methods captured the original signal well, but the EMD method appeared to result in distortion and did not fully reflect the original data information.

In summary, The EMD method was able to effectively suppress the random signal of the calibration process but failed to effectively protect the original signal characteristics of the accelerometer. The TFPF algorithm could also filter out a lot of the random signals, but the effect was less obvious. Only by combining the EMD and TFPF algorithms was the random signal effectively suppressed and the basic characteristics of the reserved signal ensured.

For the spectrum analysis of the accelerometer, the main focus was on the three states of the accelerometer calibration process. The first stage is the preparation stage, the second stage is the shock stage, and the third stage is the resonant frequency of the vibration stage (Figure 9 shows the three stages).

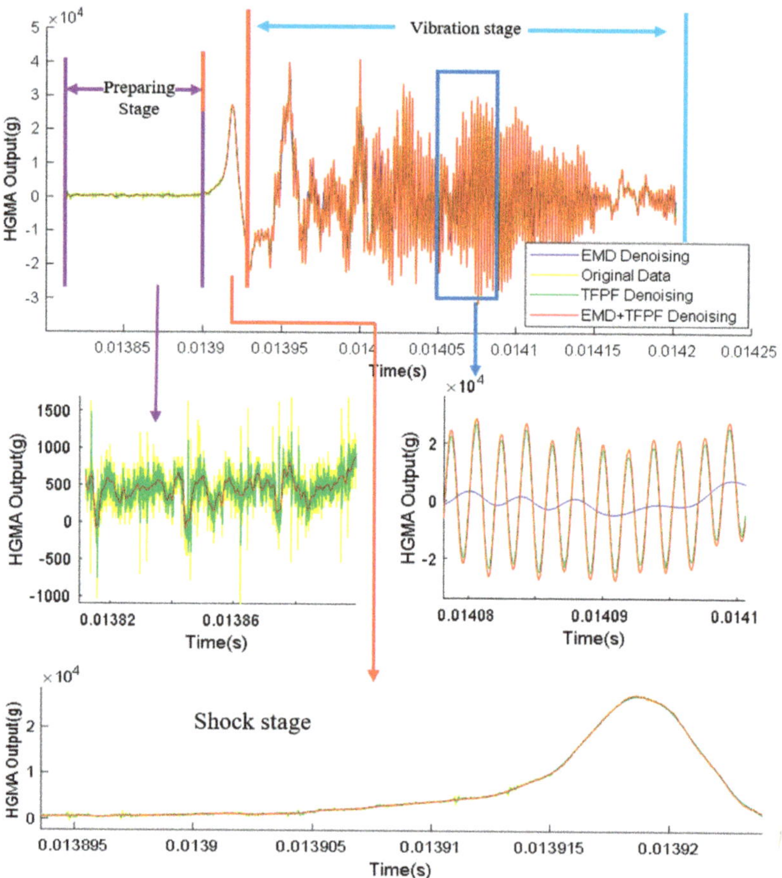

Figure 9. Noise reduction results at different stages.

The spectrum analysis results of the signals processed by the different algorithms are shown in Figure 10. For the shock stage: the frequency peak was approximately 27.1 kHz. During this stage, the amplitudes of the original signal data, the EMD, TFPF and EMD + TFPF denoising signals were 3608 g, 3541 g, 3624 g and 3610 g, respectively. Therefore, the true amplitude and frequency information of the original signal was able to be captured by the three denoising methods.

For the vibration phase: the frequency peak was approximately 525.8 kHz and the amplitudes of the original signal data, the EMD denoising result, the TFPF denoising result, and the EMD + TFPF denoising result were 5284 g, 472.6 g, 4585 g, 5310 g, respectively. The amplitude and shape of the original signal data, and the TFPF and the EMD + TFPF denoising results, were basically the same. The EMD denoising method distortion prevented accurate reflection of the frequency and amplitude information of the original data.

The results of the several denoising methods in the preparation stage were evaluated using Allan analysis of variance, as shown in Figure 11. It can be seen that the random walking and bias stability of the original data was 6.9202×10^8 (g/h), which improved to 8.102×10^7 (g/h), 8.8901×10^7 (g/h), and 1.8002×10^7 (g/h) using the EMD method, the TFPF method and the EMD + TFPF method, respectively. The results indicate that the effect of the fusion algorithm was the most obvious among the three methods.

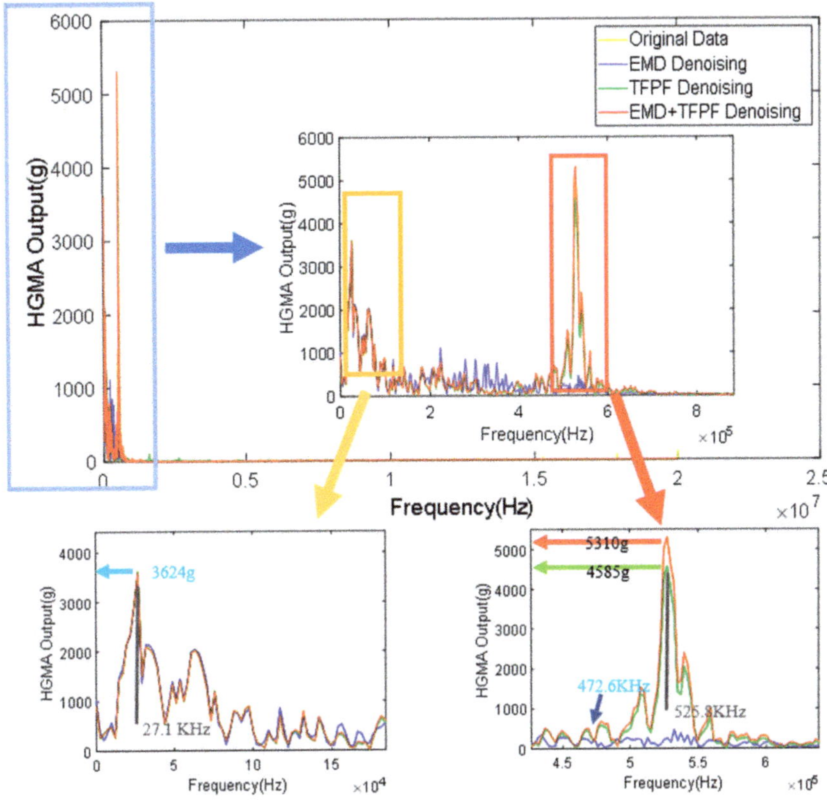

Figure 10. Frequency characteristics comparison of different denoising results.

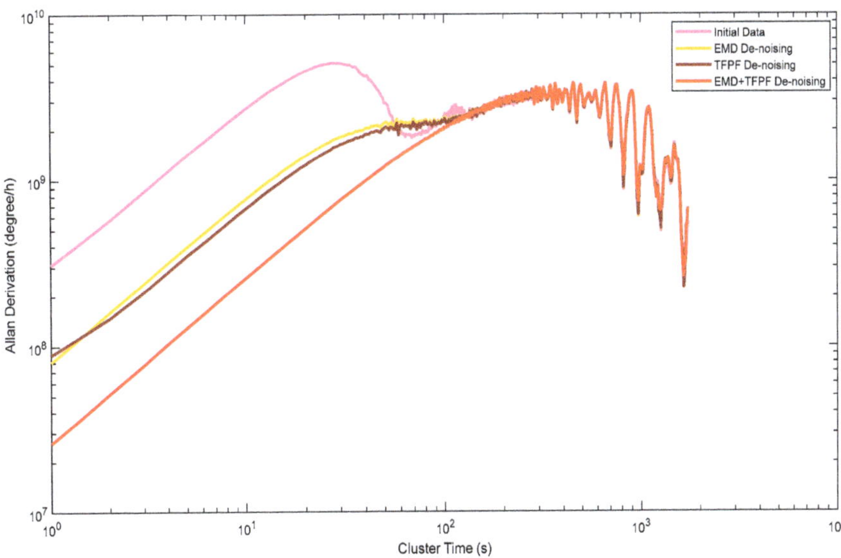

Figure 11. Allan analysis results of the denoising method during the "preparation stage".

The denoising results of the different methods at different stages are shown in Figure 12. Compared with the EMD method and the TFPF method, the EMD + TFPF method worked best in the three stages.

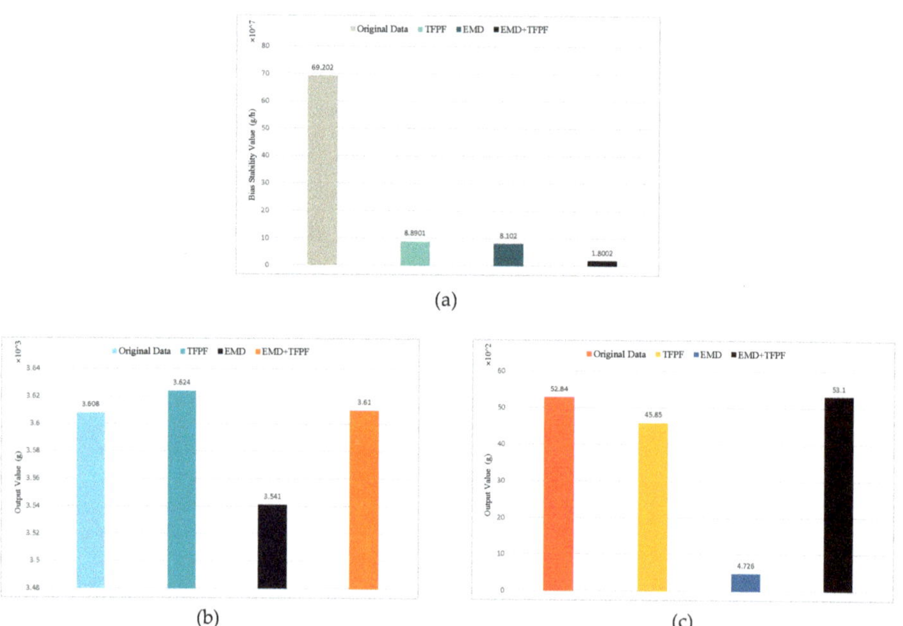

(a)

(b)

(c)

Figure 12. Noise reduction effect of the three algorithms at different stages. (**a**) is the increase in the bias stability value in the preparation stage, (**b**,**c**) are the optimization results in the shock stage and the vibration stage, respectively.

It was calculated that, in the preparation stage, the noise of the original signal was reduced by 88.3%, 87.2% and 97.4% using the EMD, TFPF, and EMD + TFPF denoising methods, respectively. In the shock stage, the errors from the original three denoising methods of EMD, TFPF, EMD + TFPF were 0.94%, 0.44%, and 0.055%, respectively. In the vibration stage, the errors for the EMD, TFPF, and EMD + TFPF methods were 91.06%, 13.23%, and 0.49%, respectively.

The results for the preparation stage showed that the EMD + TFPF method achieved the best denoising effect. In the shock stage and the vibration stage, the EMD + TFPF method did not destroy the original calibration data, and was able to perform data processing during the calibration process.

5. Conclusions

This paper introduced the high-g accelerometer and the filtering methods used in the accelerometer calibration process. The structure of the accelerometer was analyzed using finite element analysis software. Simulation results showed that the resonant frequency of HGMA in the 1st-order mode was 408 kHz. The 2nd-order resonant frequency was 667 kHz and the gap with the 1st-order mode was 260 kHz, so the coupling movement between the two modes was small, and the HGMA had better linearity. For noise reduction of the accelerometer calibration, the EMD + TFPF method was proposed. The method was mainly divided into the following steps: First, the IMF component of the high frequency band after EMD decomposition was removed; second, the IFPF algorithm was used to process the IMF component of the middle frequency band; third, the IMF component of the low frequency band was reserved; finally, the signal was reconstructed. The experimental results showed that the EMD + TFPF method was able to not only achieve a significant filtering effect, but also to retain the original signal to a large extent.

Author Contributions: Funding acquisition, C.S.; investigation, Y.L. and C.S.; software, Y.C. and K.L.; validation, Y.C. and K.L.; writing—original draft, C.W. All authors have read and agreed to the published version of the manuscript.

Funding: This work is supported by the National Natural Science Foundation of China (No. 51705477, and No. 61973281), the China Postdoctoral Innovative Talents Support Program (BX20180276), the China Postdoctoral Science Foundation (2018M641684), and the Pre-Research Field Foundation of the Equipment Development Department of China No. 61405170104. The research is also supported by the program for the Top Young Academic Leaders of Higher Learning Institutions of Shanxi, the Fund Program for the Scientific Activities of Selected Returned Overseas Professionals in Shanxi Province, the Shanxi Province Science Foundation for Youths (No. 201801D221195), the Key Research and Development (R&D) Projects of Shanxi Province (No. 201903D111005), the Aeronautical Science Foundation of China (No. 2019080U0002), and the Fund for Shanxi "1331 Project" Key Subjects Construction.

Data Availability Statement: The data used to support the findings of this study are available from the corresponding author upon request.

Conflicts of Interest: The authors declare no conflict interest.

References

1. Huang, H.; Shi, R.; Zhou, J.; Yang, Y.; Song, R.; Chen, J.; Wu, G.; Zhang, J. Attitude Determination Method Integrating Square-Root Cubature Kalman Filter with Expectation-Maximization for Inertial Navigation System Applied to Underwater Glider. *Rev. Sci. Instrum.* **2019**, *90*, 095001. [CrossRef] [PubMed]
2. Shen, C.; Liu, X.; Cao, H.; Zhou, Y.; Liu, J.; Tang, J.; Guo, X.; Huang, H.; Chen, X. Brain-like Navigation Scheme Based on MEMS-INS and Place Recognition. *Appl. Sci.* **2019**, *9*, 1708. [CrossRef]
3. Shen, C.; Zhang, Y.; Tang, J.; Cao, H.; Liu, J. Dual-Optimization for a MEMS-INS/GPS System during GPS Outages Based on the Cubature Kalman Filter and Neural Networks. *Mech. Syst. Signal Proc.* **2019**, *133*, 106222. [CrossRef]
4. Wang, Q.; Vogt, H. With PECVD Deposited Poly-SiGe and Poly-Ge Forming Contacts Between MEMS and Electronics. *J. Electon. Mater.* **2019**, *48*, 7360–7365. [CrossRef]
5. Cao, H.; Xue, R.; Cai, Q.; Gao, J.; Zhao, R.; Shi, Y.; Huang, K.; Shao, X.; Shen, C. Design and Experiment for Dual-Mass MEMS Gyroscope Sensing Closed-Loop System. *IEEE Access* **2020**, *8*, 48074–48087. [CrossRef]
6. Cao, H.; Zhang, Y.; Shen, C.; Liu, Y.; Wang, X. Temperature Energy Influence Compensation for MEMS Vibration Gyroscope Based on RBF NN-GA-KF Method. *Shock Vib.* **2018**, *2018*, 2830686. [CrossRef]
7. Shi, Y.; Wen, X.; Zhao, Y.; Zhao, R.; Cao, H.; Liu, J. Investigation and Experiment of High Shock Packaging Technology for High-G MEMS Accelerometer. *IEEE Sens. J.* **2020**, *20*, 9029–9037. [CrossRef]
8. Shi, Y.; Wang, Y.; Feng, H.; Zhao, R.; Cao, H.; Liu, J. Design, Fabrication and Test of a Low Range Capacitive Accelerometer with Anti-Overload Characteristics. *IEEE Access* **2020**, *8*, 26085–26093. [CrossRef]
9. Zhang, H.; Wei, X.; Ding, Y.; Jiang, Z.; Ren, J. A Low Noise Capacitive MEMS Accelerometer with Anti-Spring Structure. *Sens. Actuators A Phys.* **2019**, *296*, 79–86. [CrossRef]
10. Rao, K.; Wei, X.; Zhang, S.; Zhang, M.; Hu, C.; Liu, H.; Tu, L.-C. A MEMS Micro-g Capacitive Accelerometer Based on through-Silicon-Wafer-Etching Process. *Micromachines* **2019**, *10*, 380. [CrossRef]
11. Kamada, Y.; Isobe, A.; Oshima, T.; Furubayashi, Y.; Ido, T.; Sekiguchi, T. Capacitive MEMS Accelerometer with Perforated and Electrically Separated Mass Structure for Low Noise and Low Power. *J. Microelectromech. Syst.* **2019**, *28*, 401–408. [CrossRef]
12. Yeh, C.Y.; Huang, J.T.; Tseng, S.-H.; Wu, P.-C.; Tsai, H.-H.; Jualng, Y.-Z. A Low-Power Monolithic Three-Axis Accelerometer with Automatically Sensor Offset Compensated and Interface Circuit. *Microelectron. J.* **2019**, *86*, 150–160. [CrossRef]
13. Utz, A.; Walk, C.; Stanitzki, A.; Mokhtari, M.; Kraft, M.; Kokozinski, R. A High-Precision and High-Bandwidth MEMS-Based Capacitive Accelerometer. *IEEE Sens. J.* **2018**, *18*, 6533–6539. [CrossRef]
14. Liu, Y.; Zhang, L.; Wang, B. A Low Power Accelerometer System with Hybrid Signal Output. *IEICE Electron. Express* **2018**, *15*, 20171091. [CrossRef]
15. Najafi, A.; Keighobadi, J. Full-State-Feedback, Fuzzy Type I and Fuzzy Type II Control of MEMS Accelerometer. *J. Mech. Sci. Technol.* **2018**, *32*, 793–798. [CrossRef]
16. Edalafar, F.; Azimi, S.; Qureshi, A.Q.A.; Yaghootkar, B.; Keast, A.; Friedrich, W.; Leung, A.M.; Bahreyni, B. A Wideband, Low-Noise Accelerometer for Sonar Wave Detection. *IEEE Sens. J.* **2018**, *18*, 508–516. [CrossRef]
17. Yan, Z.; Hou, B.; Zhang, J.; Shen, C.; Shi, Y.; Tang, J.; Cao, H.; Liu, J. MEMS Accelerometer Calibration Denoising Method for Hopkinson Bar System Based on LMD-SE-TFPF. *IEEE Access* **2019**, *7*, 113901–113915. [CrossRef]

18. Guo, H.; Hong, H. Research on Filtering Algorithm of MEMS Gyroscope Based on Information Fusion. *Sensors* **2019**, *19*, 3552. [CrossRef]
19. Zou, Y.; Chen, Y.; Liu, P. Refactoring and Optimization of Bridge Dynamic Displacement Based on Ensemble Empirical Mode Decomposition. *Sensors* **2019**, *19*, 3125. [CrossRef]
20. Ding, H.; Wang, Y.; Yang, Z.; Pfeiffer, O. Nonlinear Blind Source Separation and Fault Feature Extraction Method for Mining Machine Diagnosis. *Appl. Sci.* **2019**, *9*, 1852. [CrossRef]
21. Shen, C.; Song, R.; Li, J.; Zhang, X.; Tang, J.; Shi, Y.; Liu, J.; Cao, H. Temperature Drift Modeling of MEMS Gyroscope Based on Genetic-Elman Neural Network. *Mech. Syst. Signal Proc.* **2016**, *72–73*, 897–905. [CrossRef]
22. Jiang, C.; Chen, S.; Chen, Y.; Zhang, B.; Feng, Z.; Zhou, H.; Bo, Y. A MEMS IMU De-Noising Method Using Long Short Term Memory Recurrent Neural Networks (LSTM-RNN). *Sensors* **2018**, *18*, 3470. [CrossRef] [PubMed]
23. Lu, Q.; Pang, L.; Huang, H.; Shen, C.; Cao, H.; Shi, Y.; Liu, J. High-G Calibration Denoising Method for High-G MEMS Accelerometer Based on EMD and Wavelet Threshold. *Micromachines* **2019**, *10*, 134. [CrossRef] [PubMed]
24. Zhu, M.; Pang, L.; Xiao, Z.; Shen, C.; Cao, H.; Shi, Y.; Liu, J. Temperature Drift Compensation for High-G MEMS Accelerometer Based on RBF NN Improved Method. *Appl. Sci.* **2019**, *9*, 695. [CrossRef]
25. Zhang, T.; Liu, H.; Feng, L.; Wang, X.; Zhang, Y. Noise Suppression of a Micro-Grating Accelerometer Based on the Dual Modulation Method. *Appl. Opt.* **2017**, *56*, 10003–10008. [CrossRef]
26. Mokhtari, E.; Elhabiby, M.; Sideris, M.G. Wavelet Spectral Techniques for Error Mitigation in the Superconductive Angular Accelerometer Output of a Gravity Gradiometer System. *IEEE Sens. J.* **2017**, *17*, 3782–3793. [CrossRef]
27. He, J.; Bai, S.; Wang, X. An Unobtrusive Fall Detection and Alerting System Based on Kalman Filter and Bayes Network Classifier. *Sensors* **2017**, *17*, 1393. [CrossRef]
28. Abbasi-Kesbi, R.; Nikfarjam, A. Denoising MEMS Accelerometer Sensors Based on L2-Norm Total Variation Algorithm. *Electron. Lett.* **2017**, *53*, 322–323. [CrossRef]
29. Kou, Y.; Kou, Z.; Yang, J. Hybrid De-Noising Algorithm Based on FLP, Grey AGO and LWT. *J. Grey Syst.* **2017**, *29*, 125.
30. Shen, C.; Yang, J.; Tang, J.; Liu, J.; Cao, H. Note: Parallel Processing Algorithm of Temperature and Noise Error for Micro-Electro-Mechanical System Gyroscope Based on Variational Mode Decomposition and Augmented Nonlinear Differentiator. *Rev. Sci. Instrum.* **2018**, *89*, 076107. [CrossRef]
31. Wang, Z.; Wang, J.; Cai, W.; Zhou, J.; Du, W.; Wang, J.; He, G.; He, H. Application of an Improved Ensemble Local Mean Decomposition Method for Gearbox Composite Fault Diagnosis. *Complexity* **2019**, *2019*, 1564243. [CrossRef]
32. Cao, H.; Li, H.; Liu, J.; Shi, Y.; Tang, J.; Shen, C. An Improved Interface and Noise Analysis of a Turning Fork Microgyroscope Structure. *Mech. Syst. Signal Proc.* **2016**, *70–71*, 1209–1220. [CrossRef]
33. Wang, Z.; Du, W.; Wang, J.; Zhou, J.; Han, X.; Zhang, Z.; Huang, L. Research and Application of Improved Adaptive MOMEDA Fault Diagnosis Method. *Measurement* **2019**, *140*, 63–75. [CrossRef]
34. Cao, H.; Liu, Y.; Zhang, Y.; Shao, X.; Gao, J.; Huang, K.; Shi, Y.; Tang, J.; Shen, C.; Liu, J. Design and Experiment of Dual-Mass MEMS Gyroscope Sense Closed System Based on Bipole Compensation Method. *IEEE Access* **2019**, *7*, 49111–49124. [CrossRef]
35. Cai, Q.; Zhao, F.; Kang, Q.; Luo, Z.; Hu, D.; Liu, J.; Cao, H. A Novel Parallel Processing Model for Noise Reduction and Temperature Compensation of MEMS Gyroscope. *Micromachines* **2021**, *12*, 1285. [CrossRef]
36. Shen, C.; Cao, H.; Li, J.; Tang, J.; Zhang, X.; Shi, Y.; Yang, W.; Liu, J. Hybrid De-Noising Approach for Fiber Optic Gyroscopes Combining Improved Empirical Mode Decomposition and Forward Linear Prediction Algorithms. *Rev. Sci. Instrum.* **2016**, *87*, 033305. [CrossRef]
37. Cao, H.; Liu, Y.; Liu, L.; Wang, X. Humidity Drift Modeling and Compensation of MEMS Gyroscope Based on IAWTD-CSVM-EEMD Algorithms. *IEEE Access* **2021**, *9*, 95686–95701. [CrossRef]
38. Ma, T.; Cao, H.; Shen, C. A Temperature Error Parallel Processing Model for MEMS Gyroscope Based on a Novel Fusion Algorithm. *Electronics* **2020**, *9*, 499. [CrossRef]
39. Cao, H.; Wei, W.; Liu, L.; Ma, T.; Zhang, Z.; Zhang, W.; Shen, C.; Duan, X. A Temperature Compensation Approach for Dual-Mass MEMS Gyroscope Based on PE-LCD and ANFIS. *IEEE Access* **2021**, *9*, 95180–95193. [CrossRef]
40. Ma, T.; Li, Z.; Cao, H.; Shen, C.; Wang, Z. A Parallel Denoising Model for Dual-Mass MEMS Gyroscope Based on PE-ITD and SA-ELM. *IEEE Access* **2019**, *7*, 169979–169991. [CrossRef]
41. Shi, Y.; Yang, Z.; Ma, Z.; Cao, H.; Kou, Z.; Zhi, D.; Chen, Y.; Feng, H.; Liu, J. The Development of a Dual-Warhead Impact System for Dynamic Linearity Measurement of a High-g Micro-Electro-Mechanical-Systems (MEMS) Accelerometer. *Sensors* **2016**, *16*, 840. [CrossRef] [PubMed]
42. Shi, Y.; Zhao, Y.; Feng, H.; Cao, H.; Tang, J.; Li, J.; Zhao, R.; Liu, J. Design, Fabrication and Calibration of a High-G MEMS Accelerometer. *Sens. Actuator A Phys.* **2018**, *279*, 733–742. [CrossRef]
43. Liao, M.; Guo, Y.; Qin, Y.; Wang, Y. The Application of EMD in Activity Recognition Based on a Single Triaxial Accelerometer. *Bio-Med. Mater. Eng.* **2015**, *26*, S1533–S1539. [CrossRef] [PubMed]
44. Shen, C.; Li, J.; Zhang, X.; Shi, Y.; Tang, J.; Cao, H.; Liu, J. A Noise Reduction Method for Dual-Mass Micro-Electromechanical Gyroscopes Based on Sample Entropy Empirical Mode Decomposition and Time-Frequency Peak Filtering. *Sensors* **2016**, *16*, 796. [CrossRef] [PubMed]
45. Liu, Y.; Yue, L.; Lin, H.; Ma, H. An Amplitude-Preserved Time–Frequency Peak Filtering Based on Empirical Mode Decomposition for Seismic Random Noise Reduction. *IEEE Geosci. Remote Sens. Letters* **2013**, *11*, 896–900. [CrossRef]

46. Lin, T.; Zhang, Y.; Muller-Petke, M. Random Noise Suppression of Magnetic Resonance Sounding Oscillating Signal by Combining Empirical Mode Decomposition and Time-Frequency Peak Filtering. *IEEE Access* **2019**, *7*, 79917–79926. [CrossRef]
47. Oppenheim, A.V.; Willsky, A.S.; Nawab, S.H. *Signals and Systems*, 2nd ed.; Liu, S., Ed.; Electronic Industry Press: Beijing, China, 2015.
48. Xiong, M.; Li, Y.; Wu, N. Random-Noise Attenuation for Seismic Data by Local Parallel Radial-Trace TFPF. *IEEE Trans. Geosci. Remote Sens.* **2013**, *52*, 4025–4031. [CrossRef]

micromachines

Article

Research on Optical Fiber Ring Resonator Q Value and Coupling Efficiency Optimization

Shengkun Li *, Xiaowen Tian and Sining Tian

School of Instrument and Electronics, North University of China, Taiyuan 030051, China
* Correspondence: lishengkun@nuc.edu.cn

Abstract: The coupling efficiency of the fiber ring resonator has an important influence on the scale factor of the resonant fiber gyroscope. In order to improve the scale factor of the gyroscope, the coupling efficiency of the fiber ring resonator and its influential factors on the scale factor of the gyroscope are analyzed and tested. The results show that the coupling efficiency is affected by both the splitting ratio of the coupler and the loss in the cavity. When the coupling efficiency approaches 0.75 at the under-coupling state, the scaling factor of the gyroscope is the highest. This provides a theoretical reference and an experimental basis for the enhancement of the scaling factor of the resonant fiber gyroscope with the fiber ring resonator as the sensitive unit, providing options for multiple applications such as sea, land, sky and space.

Keywords: resonant fiber gyroscope; fiber ring resonator; coupling efficiency; scale factor

Citation: Li, S.; Tian, X.; Tian, S. Research on Optical Fiber Ring Resonator Q Value and Coupling Efficiency Optimization. *Micromachines* **2023**, *14*, 1680. https://doi.org/10.3390/mi14091680

Academic Editor: Giancarlo C. Righini

Received: 6 July 2023
Revised: 21 August 2023
Accepted: 25 August 2023
Published: 28 August 2023

1. Introduction

Inertial navigation is a self-service navigation system that does not need external information that radiates energy to the outside world. It is widely used in national defense [1], civil and commercial fields due to its advantages of long working hours, high stability and fast data update rate [2]. A gyroscope is the core component of an inertial navigation system [3], which is used to measure the angular velocity of the vector in the inertial space and the attitude of the vector [4]; the performance of the gyroscope plays a decisive role in the performance of the whole inertial navigation system [5]. Resonant gyroscopes use the resonant properties of vibration to measure rotation [6], and resonant fiber optic gyroscope (RFOG), a novel opto-electronic hybrid integrated sensor [7,8], measure the rotational angular velocity of the system by measuring the resonance frequency difference of clockwise (CW) and counterclockwise (CCW) light caused by the optical Sagnac effect. Due to its advantages of high precision, small size and no moving parts, the resonant optical gyroscope has drawn much attention [9]. Like ordinary mechanical gyroscopes, optical gyroscopes are also affected by coupling effects, temperature [10] and other external noise [11]. The optical gyroscope is favored in various application fields such as sea, land, sky and space because of its internal all-solid-state optical path, no rotor wear and no moving parts, which greatly improve the mechanical properties of inertial devices such as anti-vibration and shock resistance. Resonant fiber optic gyroscopes, as a kind of optical gyroscope, compared with interferometric fiber optic gyroscopes, only need to use a 2×2 coupler and pigtail fusion to form a resonant cavity, avoiding the temperature drift caused by the Shupe effect in the long optical fiber sensitive annulus At the same time, the miniaturization also shows its unique advantages and engineering application potential. A fiber ring resonator works as the core sensitive unit of the RFOG; it can be simply formed by splicing coupler pigtails. The low insertion loss allows light energy to be stored in the fiber core efficiently [12].

Although the fiber ring resonator (FRR) can improve the detection sensitivity of the gyroscope by obtaining a higher Q value, the effect of coupling efficiency on the gyroscopic scaling factor has not yet been fully analyzed theoretically and experimentally. Thus, in order to ensure the stability and accuracy of the gyroscope, it is necessary to reduce its temperature drift and coupling effects [13]. There are more research works on the direction of temperature drift compensation, and the combination of temperature compensation algorithms can effectively reduce the drift amount [14]. However, the cross-coupling effect [15,16] between multiple axes of different gyroscopes has some differences, and it is important to deeply explore and analyze its sources and mechanisms. The work in this paper focuses on the analysis of the coupling states of FRR, which has three coupling states, namely, the states of under-coupling, over-coupling and critical coupling [17]. Experiments are conducted under different coupling states to investigate in depth the effects of different coupling states on the accuracy of gyroscopes and to provide an experimental basis for the subsequent related calibration and compensation.

The influence factors of the three coupled states of the fiber ring resonator are studied in this paper. The influence of the coupled state on the detection accuracy of the resonant optical gyroscope is studied through theoretical analysis, modeling simulation and experimental testing. The experimental results show that the coupling efficiency of the resonant cavity is mainly influenced by the cavity loss and the split ratio of coupling. When the coupling efficiency approaches 0.75, the gyroscope has the highest scale factor and is in the under-coupling state. The cavity length is 3 m, and, when the split ratio is 2/98, a resonator with Q value of 2.3×10^8 and coupling efficiency of 0.7334 is obtained, with the highest top scale factor of 118.2876 mV/°/s.

2. The Analysis of Principle

2.1. Coupling Efficiency (ρ)

As the core component of the fiber ring resonator, the fiber coupler uses the evanescent field coupling of the light waves to realize the splitting and combining between the adjacent fibers [18].

Figure 1 shows the fiber ring resonator. E_1, E_2, E_3 and E_4 are the intensity of the light field propagating in ports 1, 2, 3 and 4. Light enters the fiber coupler from port 1, and part of it is output from port 2 directly. The other part of the light is coupled into the cavity via port 4 and continues to propagate through the coupling area for one more revolution before entering the coupler through port 3 again. The light entering port 3 continues to propagate through port 4, while the other portion enters the coupling region to exit port 2. E_1, E_2, E_3 and E_4 satisfy the following relationship [19]:

$$E_2 = \gamma E_1 + jt E_3 \tag{1}$$

$$E_4 = \gamma E_3 + jt E_1 \tag{2}$$

$$\gamma^2 + t^2 = 1 \tag{3}$$

γ and t are the coupling and transmission coefficients, respectively, and j is the imaginary unit.

The loop light intensity transfer factor τ and transmission phase shift φ are used to describe the propagation of light in the fiber ring resonator:

$$E_2 = \tau \exp(i\varphi) E_4 \tag{4}$$

where $\varphi = \kappa L$, $\kappa = 2\pi n / \lambda$, n is the effective refractive index of the fiber, λ is the wavelength of the incident light in vacuum and L is the cavity length of the fiber ring resonator.

From (1)–(4), the relationship between E_1 and E_2 can be expressed:

$$\frac{E_2}{E_1} = \exp[i(\pi + \varphi)] \frac{\tau - \gamma \exp(-i\varphi)}{1 - \gamma\tau \exp(-i\varphi)} \tag{5}$$

The transmission light can be obtained by the square of Formula (5):

$$T = \left|\frac{E_2}{E_1}\right| = \frac{\tau^2 - 2\gamma\tau\cos\varphi + \gamma^2}{1 - 2\gamma\tau\cos\varphi + \gamma^2\tau^2} \tag{6}$$

Based on the above calculations, the coupling efficiency can be calculated as follows [9]:

$$\rho = 1 - \left(\frac{\gamma - \tau}{1 - \gamma\tau}\right)^2 \tag{7}$$

The coupling state is shown in Figure 2. When $\gamma = \tau$ and $T = 0$, the output light intensity of E_2 is zero. The optical energy loss and replenishment in the resonator reach dynamic balance. The FRR is in the critical coupling state and the coupling efficiency is the highest. In this case, the light coupled to the FRR by E_1 is equal to the light loss in the cavity (including the insertion loss, the transmission loss in the fiber loop and the loss in the splice). When $\gamma > \tau$, the optical energy in the cavity is replenished more than the loss, the cavity is in an over-coupled state. Large full width at half maximum (FWHM) results in low Q factor. When $\gamma < \tau$, the cavity light energy is less than the loss of replenishment, the resonator is in a state of under-coupling, the full width at half maximum is narrow and the quality factor is high.

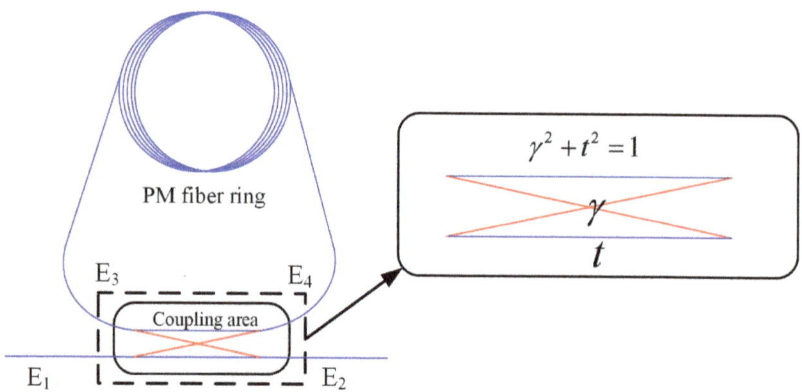

Figure 1. The structure of FRR.

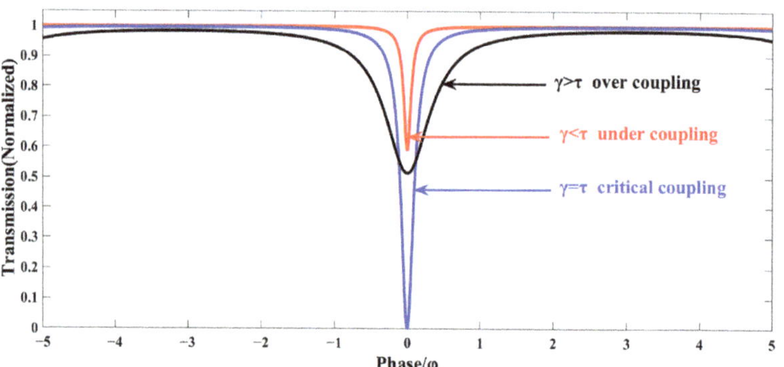

Figure 2. Simulation of coupling state.

The relationship between the splitting ratio and the loss on the resonance depth is shown in Figure 3. The coupling state of the resonator is affected by the different matching relationship between the cavity loss and the splitting ratio of the coupler. When $\gamma = \tau$, the cavity coupling efficiency is the highest.

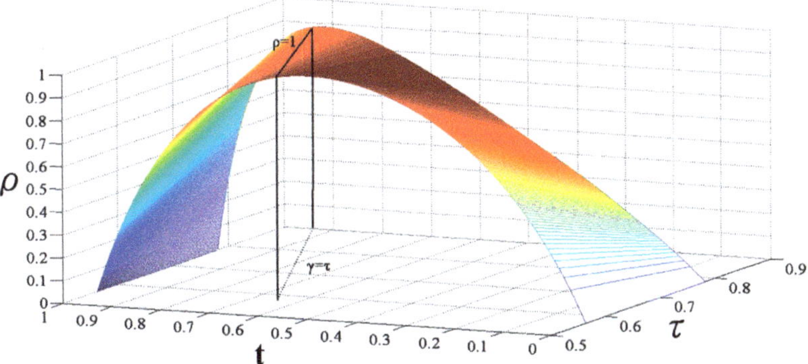

Figure 3. The influence of splitting ratio and loss on coupling efficiency.

2.2. The Scale Factor of the Gyroscope

In the RFOG signal detection system, the triangular wave phase modulation technology is used to modulate the light emitted by the laser. If the output frequency of the modulated laser turns from f_0 to $f_0 \pm \Delta f$, the gyroscope output [17,19] is

$$I_{out} = I_{in} \left(\frac{r^2 + a^2 - 2ra \cos\left(\frac{2\pi(f_0 + 2\Delta f)}{FSR}\right)}{1 + r^2 a^2 - 2ra \cos\left(\frac{2\pi(f_0 + 2\Delta f)}{FSR}\right)} - \frac{r^2 + a^2 - 2ra \cos\left(\frac{2\pi(f_0 - 2\Delta f)}{FSR}\right)}{1 + r^2 a^2 - 2ra \cos\left(\frac{2\pi(f_0 - 2\Delta f)}{FSR}\right)} \right) \tag{8}$$

We demodulate the slope of the curve under the condition of a certain free spectral range (*FSR*) of the resonance spectrum, which is the effect of fiber ring resonator coupling efficiency on gyroscope scale factor:

$$\kappa = \left. \frac{dI_{out}}{\Delta f} \right|_{\Delta f = 0} \tag{9}$$

From Equations (8) and (9), it can be known that, when the maximum value is taken, the following relation is obtained:

$$r = \frac{2\tau + 1}{\tau + 2} \tag{10}$$

In this case, the coupling efficiency $h = 0.75$, that is, when r and τ satisfy the relationship shown in Formula (10), the slope of the RFOG demodulation curve is the largest, which means that the gyroscope has the highest scale factor. In this case, the fiber ring resonator is in an under-coupled state and the coupling efficiency is 0.75.

The proper match between splitting ratio and the cavity loss in the coupler constitutes a resonant cavity with a coupling efficiency of 0.75. Losses in FRR include insertion loss, transmission loss in fiber and splice loss. Insertion loss and fiber transmission loss (0.15 dB/km) are very low compared to splice loss, so splice loss constitutes the major loss in the cavity. The splice loss was tested in the experiment.

3. The Experimental Test

3.1. Melting Point Loss Experiment

In order to obtain the wastage of the melting point, we cut off a polarization-maintaining optical fiber, and then welded it in the same welding environment and parameters. Optical fiber lasers were connected at one end, and the other end accessed ab optical power meter and read the light power before and after welding. We tested many times to obtain the average; the results are shown in Figure 4. The loss caused by each fusion point was about 0.07 dB, and the loss caused by the two melting points was about 0.14 dB.

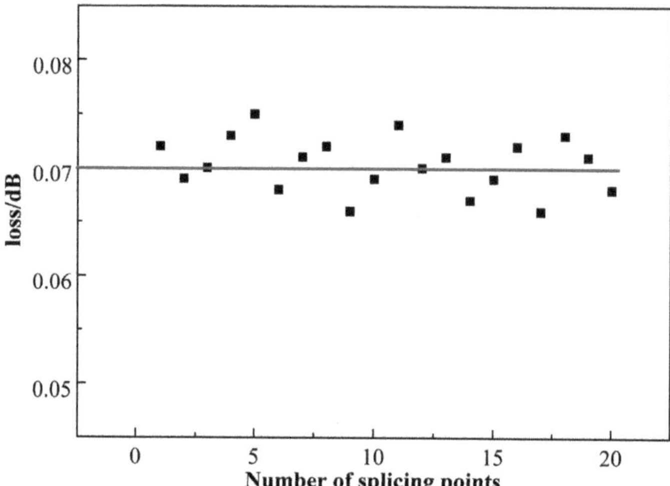

Figure 4. Loss in splicing points.

3.2. Optical Fiber Ring Resonator Test

In the resonant spectrum test system, as shown in Figure 5a, a narrow line-width laser (100 Hz) with a wavelength of 1550 nm was adopted as a laser light source, the frequency sweep coefficient was 15 MHz/V, the signal generator generated a triangle wave as the laser scanning signal and the resonance line, after photoelectric detection in photoelectric conversion, showed the form of the voltage signal on the oscilloscope. The definition of coupling efficiency is $h = \frac{I_{max} - I_{min}}{I_{max}}$, where I_{max} is the maximum value of the resonance spectrum, I_{min} is the minimum value of the resonance spectrum and $I_{max} - I_{min}$ is the amplitude range of the resonance spectrum line. The full width at half maximum (FWHM) method was adopted to calculate the Q value. The FRR resonance state was stimulated under the frequency sweep mode, sampling resonance on the scope and scanning the data of voltage. Origin software was used to calculate the scanning signal voltage difference ΔV corresponding to the FWHM of the resonance line. The laser scanning coefficient was 15 MHz/V and the corresponding frequency of light wave at the output center wavelength of 1550 nm was f_0 = 193 THz. The Q value of the resonant cavity can be calculated by the formula $Q = \frac{f_0}{FWHM} = \frac{f_0}{\Delta V \times 15 \text{ MHz/V}}$ [20]. The coupling efficiency and the Q value of the FRR are shown in Figure 5b.

In the experiment, five kinds of optical split ratio couplers (1/9, 5/95, 2/98, 1/99 and 0.1/99.9) were combined into fiber ring resonators with five cavity lengths (3 m, 5 m, 10 m, 21 m and 50 m), and the system shown in Figure 5 was used for testing and comparison with the simulation results. The results are shown in Figure 6a,b. The Q value of the resonant cavity shows an increasing trend with decrease in the splitting ratio and increase in the cavity length. The coupling efficiency is influenced by the splitting ratio of couplers and the loss in the cavity, showing a trend of an initial rise and then a decrease. This is because a high splitting ratio means energy supply is greater than the loss in the cavity, which is in the over-coupling state; additionally, the coupling efficiency and Q value are low. With further decrease in the splitting ratio, the complementary and dissipation of light energy in the cavity gradually reached dynamic equilibrium. The coupling efficiency and the Q value were gradually increased, approaching the critical coupling state. Further reduction in the splitting ratio caused the light energy supplement in the cavity to be less than the loss. In the under-coupled state, the Q value gradually increased, and the coupling efficiency showed a downward trend. The declining green curve in Figure 6b is due to the longer length of the cavity causing a bigger transmission loss and to melt contact loss accumulation making the optical energy loss in the cavity greater than the optical energy supplement to the resonant cavity under each splitting ratio; therefore, this was in the under-coupling state. The loss cavity length was 3 m. The Q value of the resonant cavity was 2.3×10^8 while the splitting ratio of the resonator was 2/98, and the coupling efficiency was 0.7334. Therefore, in the case where the optical fiber fusion point loss is known, by adopting the appropriate splitting ratio tie-in cavity length, the highest Q value and coupling efficiency of the optical FRR, i.e., 0.75, can be theoretically obtained.

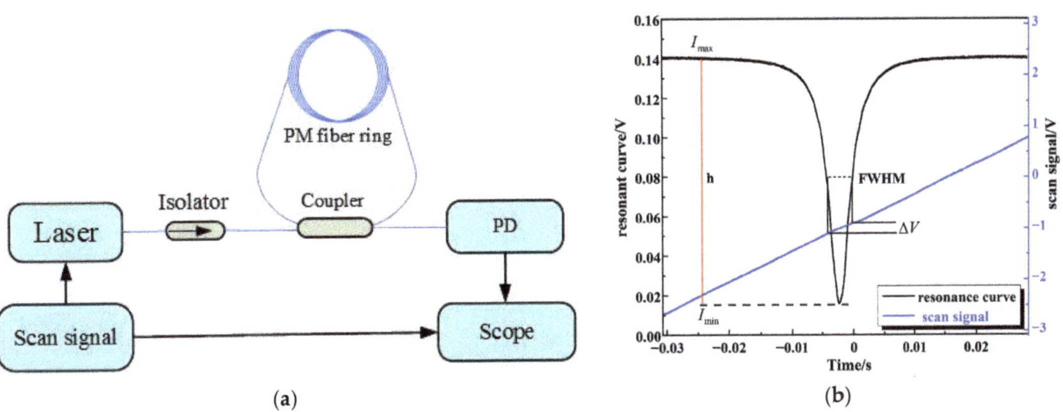

Figure 5. (**a**) Resonant spectrum test system, (**b**) coupling efficiency and Q factor calculation.

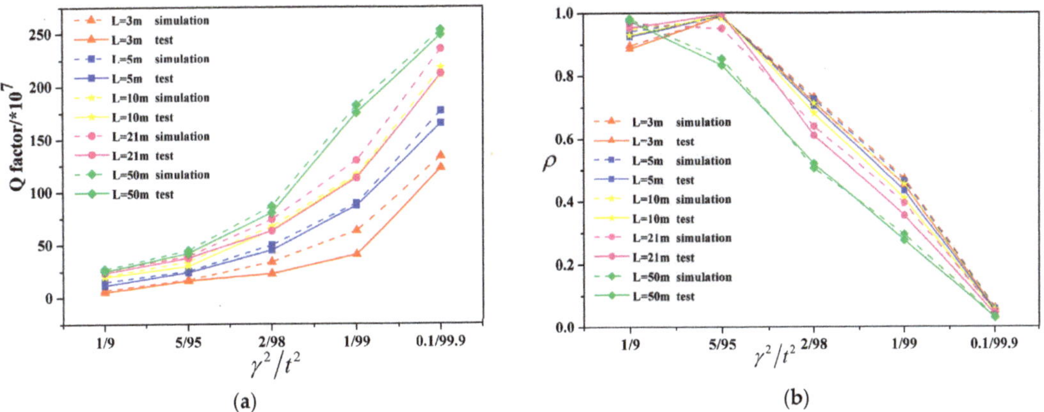

Figure 6. (**a**) Comparison of measured and simulated Q values for different lengths and splitting ratios, (**b**) comparison of measured and simulated coupling efficiency for different lengths and splitting ratios.

3.3. Resonance Fiber Optic Gyroscope Index Test

The above FRRs with different coupling efficiencies were tested in the RFOG system shown in Figure 7. Laser was emitted with stable power and wavelength of 1550 nm, and it was divided into two identical beams of light through a Y-waveguide. The two-phase modulation arms in the Y-waveguide phase modulator were used for the CW and CCW directions of optical signal modulation, in which the modulation frequencies were 335 kHz and 550 kHz, respectively. The two channels of light are resonant after the coupler; they enter the resonator cavity, pass through the circulator, are converted into an electrical signal in the photodetector and are output to the lock-in amplifier to be demodulated. In the CCW road, the output frequency of the laser is locked in as the resonant frequency of the fiber ring resonator through proportional Integration controller; the CW road is gyroscope output.

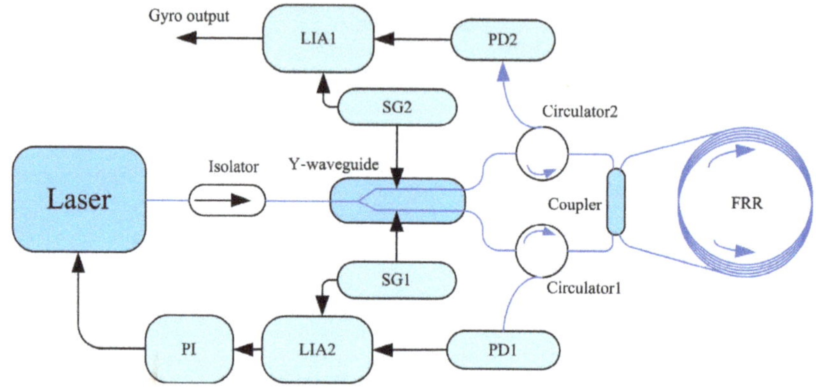

Figure 7. Schematic diagram of the ROG.

As shown in Figure 8, the prototype was fixed on the rotary table at room temperature. The frequency locking of the laser was realized by sending the frequency locking control command from the host computer, as shown in Figure 8a, to control the angular velocity output of the high-precision rotary table and to calibrate its scaling factor by using the step signals corresponding to the individual angular velocities. The output rotational angular velocity of the rotary table was controlled by the upper computer.

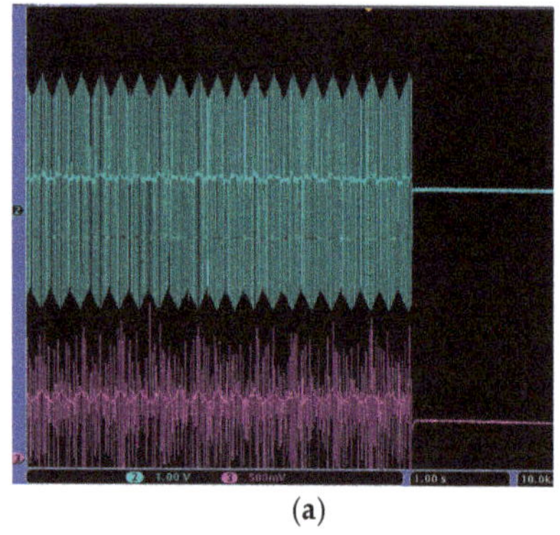
(a)

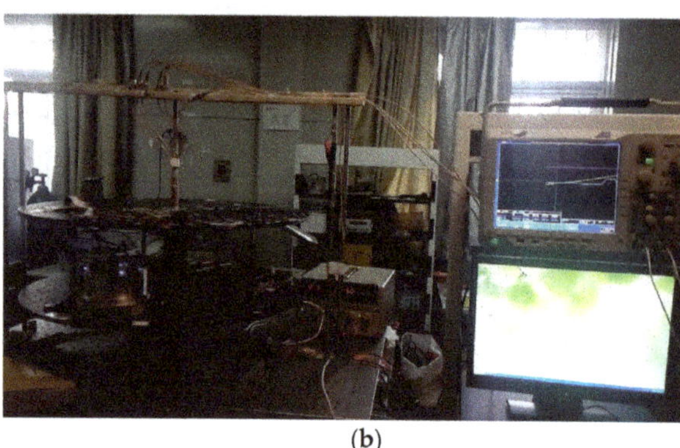
(b)

Figure 8. (a) Laser frequency locking; (b) Rotation test.

The system was fixed on the high precision turntable with the angular velocity set at about $10°/s$, $20°/s$, $30°/s$, $40°/s$ and $50°/s$ for rotation tests. We performed rotation at each angular velocity three times, taking the average of the output voltage as rotation output signal. The results in Figure 9a show the least-square linear fitting of the output voltage corresponding to each angular velocity, used to obtain the curve of angular velocity and the output voltage as shown in Figure 9b. The calculated gyroscope scale factor was $118.2876 \, mV/°/s$.

For the data in Figure 9c, the FRR with 25 parameters was tested in the gyroscope system, comparing the measured value with the theoretical calculation; the actual measurement values were basically the same as those from the theoretical calculation. With the increase in coupling efficiency, the scale factor of the gyroscope first increases and then decreases. For the scale factors between 0.641 and 0.8966, all values exceeded 100 mV/°/s, with the highest point being 118.2876 mV/°/s, corresponding to a coupling efficiency of about 0.7334. Therefore, under the current test conditions, it is necessary to control the coupling efficiency to values between 0.641 and 0.8966 to achieve a higher scale factor. There is a certain deviation between the measured value and the theoretical calculation value because the FRR has a certain error in the light ratio and the calculation error of the melting point loss.

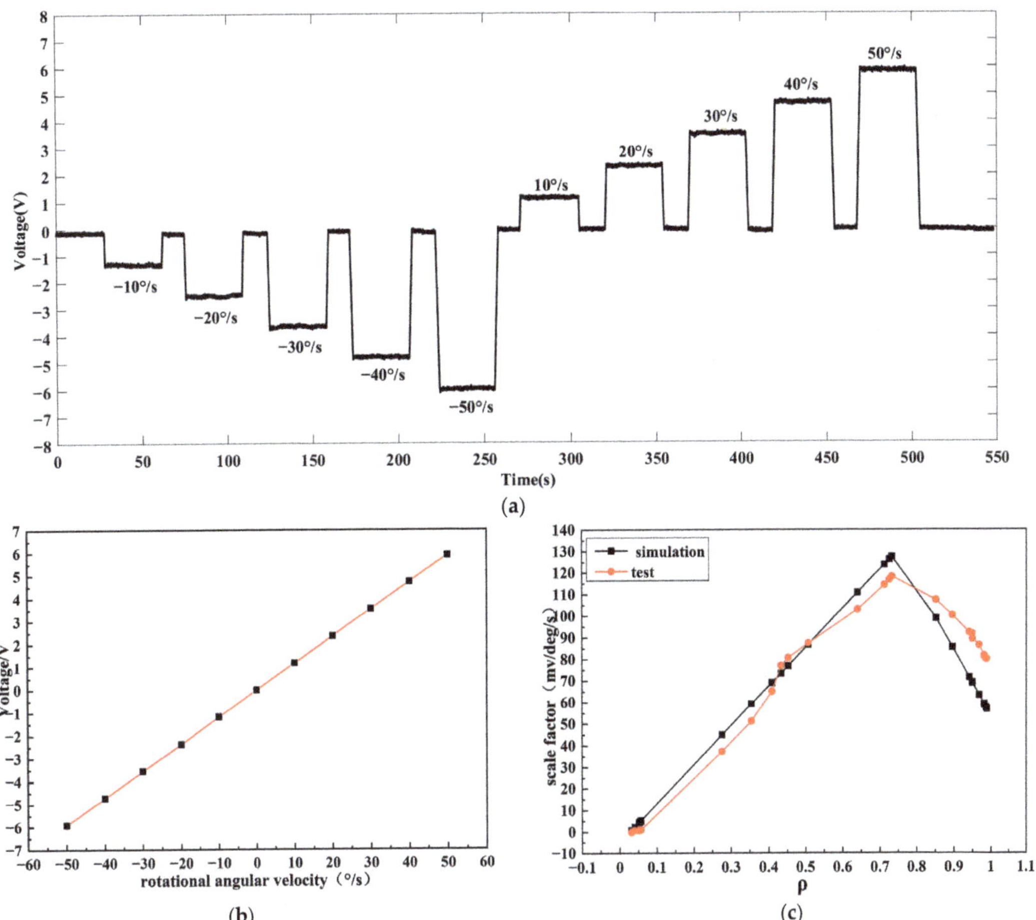

Figure 9. (a) Gyroscope signal with different rotational angular velocities; (b) fitting analysis of stair effects; (c) comparison of measured and simulated values of ROG sensitivity for the different coupling efficiency values.

4. Conclusions

In this paper, the influencing factors of coupling efficiency on FRR are modeled, and the influence of coupling efficiency on the scale factor of a resonant optical gyroscope is analyzed in detail. The Q factor and coupling efficiency test system of the FRR and the resonant fiber gyroscope system are built, and the scale factor of the gyroscope is tested with a high-precision rotating platform. Both theoretical analysis and experimental results show that the coupling efficiency of the resonant cavity is influenced by cavity loss and the splitting ratio of coupling. The scale factor of the gyroscope is the highest when the coupling efficiency approaches 0.75 in the under-coupling state. The cavity length is 3 m, and, when the splitting ratio is 2/98, a resonator with a Q value of 2.3×10^8 and a coupling efficiency of 0.7334 is obtained. In addition, the top scale factor is the highest, which is 118.2876 mV/$^\circ$/s. This paper provides theoretical and experimental support for improving the scale factor of an RFOG with a fiber ring resonator as the sensitive element. It is of great significance to improve the detection accuracy of the gyroscope's rotational angular velocity in the carrier, to enhance the initial alignment accuracy of the inertial navigation system, to obtain the carrier orientation and attitude data more accurately for the control system and to improve the accuracy and stability of the carrier's running attitude.

Author Contributions: Conceptualization, S.L.; methodology, S.L.; software, X.T.; validation, S.T.; writing, X.T.; visualization, S.T.; supervision, S.L. All authors have read and agreed to the published version of the manuscript.

Funding: This research was funded by the National Key Research and Development Program of China (No. 2022YFB3205000) and the National Natural Science Foundation of China (NSAF, Grant No. U2230206).

Data Availability Statement: Not applicable.

Conflicts of Interest: The authors declare no conflict of interest.

References

1. Duan, X.; Cao, H. Stabilized Inertial Guidance Solution for Rolling Projectile Based on Partial Strapdown Platform. *IEEE Access* **2021**, *9*, 116207–116214. [CrossRef]
2. Xie, C. Research on Enhancement Signal-to-Noise Ratio of Resonant Optical Gyroscope. Ph.D. Thesis, North University of China, Taiyuan, China, 2017.
3. Shen, C.; Xiong, Y.; Zhao, D.; Wang, C.; Cao, H.; Song, X.; Tang, J.; Liu, J. Multi-Rate Strong Tracking Square-Root Cubature Kalman Filter for MEMS-INS/GPS/Polarization Compass Integrated Navigation System. *Mech. Syst. Signal Process.* **2022**, *163*, 108146. [CrossRef]
4. Zhang, T. Research of Temperature Characteristics of Resonator Fiber Optic Gyroscope. Master's Thesis, North University of China, Taiyuan, China, 2017.
5. Cui, R.; Li, K.; Xu, X.; Xue, R.; Shen, C.; Shi, Y.; Cao, H. Design and Experiment of MEMS Solid-State Wave Gyroscope Quadrature Error Correction System. *IEEE Sens. J.* **2023**, *23*, 16645–16655. [CrossRef]
6. Guo, X.; Cui, R.; Yan, S.; Cai, Q.; Wei, W.; Shen, C.; Cao, H. Design and Experiment for N = 3 Wineglass Mode Metal Cylindrical Resonator Gyroscope Closed-Loop System. *Electronics* **2023**, *12*, 131. [CrossRef]
7. Ciminelli, C.; Dell'Olio, F.; Campanella, C.; Armenise, M. Photonic Technologies for Angular Velocity Sensing. *Adv. Opt. Photonics* **2010**, *2*, 370–404. [CrossRef]
8. Dell'Olio, F.; Indiveri, F.; Innone, F.; Dello Russo, P.; Ciminelli, C.; Armenise, M. System Test of an Optoelectronic Gyroscope Based on a High Q-Factor InP Ring Resonator. *Opt. Eng.* **2014**, *53*, 127104. [CrossRef]
9. Yan, S.; An, P.; Zheng, Y.; Li, X.; Zhao, R.; Zhang, C.; Xue, C.; Liu, J. High-Q Optical Ring Resonator Gyro Angular Rate Sensor. *Acta Photonica Sin.* **2014**, *43*, 12–17.
10. Cai, Q.; Zhao, F.; Kang, Q.; Luo, Z.; Hu, D.; Liu, J.; Cao, H. A Novel Parallel Processing Model for Noise Reduction and Temperature Compensation of MEMS Gyroscope. *Micromachines* **2021**, *12*, 1285. [CrossRef] [PubMed]
11. Li, Z.; Gu, Y.; Yang, J.; Cao, H.; Wang, G. A Noise Reduction Method for Four-Mass Vibration MEMS Gyroscope Based on ILMD and PTTFPF. *Micromachines* **2022**, *13*, 1807. [CrossRef] [PubMed]
12. Saleh, K.; Llopis, O.; Cibiel, G. Optical Scattering Induced Noise in Fiber Ring Resonators and Optoelectronic Oscillators. *J. Light. Technol.* **2013**, *31*, 1433–1446. [CrossRef]
13. Cao, H.; Cai, Q.; Zhang, Y.; Shen, C.; Shi, Y.; Liu, J. Design, Fabrication, and Experiment of a Decoupled Multi-Frame Vibration MEMS Gyroscope. *IEEE Sens. J.* **2021**, *21*, 19815–19824. [CrossRef]
14. Cao, H.; Cui, R.; Liu, W.; Ma, T.; Zhang, Z.; Shen, C.; Shi, Y. Dual Mass MEMS Gyroscope Temperature Drift Compensation Based on TFPF-MEA-BP Algorithm. *Sens. Rev.* **2021**, *41*, 162–175. [CrossRef]
15. Chen, J.; Zhao, L.; Cheng, Y.; Yan, Z.; Wang, X.; Tang, C.; Gao, F.; Yi, Z.; Zhu, M. Bandwidth-Tunable Absorption Enhancement of Visible and near-Infrared Light in Monolayer Graphene by Localized Plasmon Resonances and Their Diffraction Coupling. *Results Phys.* **2023**, *49*, 106471. [CrossRef]
16. Wu, Y.; Cai, P.; Nie, Q.; Tang, C.; Liu, F.; Zhu, M. Ultra-Narrowband, Electrically Switchable, and High-Efficiency Absorption in Monolayer Graphene Resulting from Lattice Plasmon Resonance. *Results Phys.* **2023**, *51*, 106768. [CrossRef]
17. Qian, K.; Tang, J.; Guo, H.; Liu, W.; Liu, J.; Xue, C.; Zheng, Y.; Zhang, C. Under-Coupling Whispering Gallery Mode Resonator Applied to Resonant Micro-Optic Gyroscope. *Sensors* **2017**, *17*, 100. [CrossRef] [PubMed]

MDPI
St. Alban-Anlage 66
4052 Basel
Switzerland
www.mdpi.com

Micromachines Editorial Office
E-mail: micromachines@mdpi.com
www.mdpi.com/journal/micromachines

18. Liu, Y.; Xue, C.; Zheng, H.; An, P.; Cui, X.; Lu, X.; Liu, J. Structure design and optimization of high fineness ring resonator. *Infrared Laser Eng.* **2014**, *43*, 3688–3693.
19. Ying, D.; Wang, Z.; Mao, J.; Jin, Z. An Open-Loop RFOG Based on Harmonic Division Technique to Suppress LD's Intensity Modulation Noise. *Opt. Commun.* **2016**, *378*, 10–15. [CrossRef]
20. Wang, Y. Theoretical and Experimental Study on Q Factor Enhancement in The Fiber Coupled Resonator. Master's Thesis, Harbin Institute of Technology, Harbin, China, 2013.